U0344762

COMPETITIVE BIDDINGS COLLECTION OF CHINESE ARCHITECTURE

中国建筑竞标方案集成

· 公建篇

广州佳图文化传播有限公司　主编

深圳出版发行集团
海天出版社

图书在版编目（CIP）数据

中国建筑竞标方案集成·公建篇/广州佳图文化传播有限公司主编．—深圳：海天出版社，2010.1
　（佳图建筑系列）
　ISBN 978-7-80747-758-7

　Ⅰ.中… Ⅱ.广… Ⅲ.①建筑设计-中国-图集②公共建筑-建筑设计-中国-图集 Ⅳ. TU206 TU242-64

中国版本图书馆CIP数据核字（2009）第206495号

中国建筑竞标方案集成·公建篇
ZHONGGUO JIANZHU JINGBIAO FANG'AN JICHENG·GONGJIANPIAN

出 品 人：陈锦涛
出版策划：毛世屏
责任编辑：王　颖（0755-83460593　E-mail:6021@sina.com）
责任校对：周　强
责任技编：钟愉琼
策　　划：佳图文化

出版发行　海天出版社
地　　址　深圳市彩田南路海天大厦（518033）
网　　址　www.htph.com.cn
电　　话　0755-83460137（批发）　83460397（邮购）
印　　刷　深圳市佳信达印务有限公司
版　　次　2010年1月第1版
印　　次　2010年1月第1次印刷
开　　本　889mm×1194mm　1/16
印　　张　19.5
总 定 价　520.00元（共两册）

海天版图书版权所有，侵权必究。
海天版图书凡有印装质量问题，请随时向承印厂调换。

COMPETITIVE BIDDINGS COLLECTION OF CHINESE ARCHITECTURE

公建篇

Contents

OFFICE ARCHITECTURE 办公建筑

3/Hunan People's Radio Station Technology and Office Building 湖南人民广播电台技术及办公大楼
7/Shenzhen Changfu Jinmao Mansion Design 深圳市长富金茂大厦建筑设计
13/Shenzhen Konka R&D Mansion 深圳康佳研发大厦
17/Shenzhen Diamond Tower 深圳钻石塔
19/Shenzhen Kingkey Finance Center 深圳京基金融中心
23/Jinshi International Creativity Center 金石国际创新中心
27/Shenzhen New Mansion of China Sea Oil 中国海油深圳新大厦
31/Eastern Blue Sea 东方蓝海
35/Binghai Zheshang Mansion 滨海浙商大厦
39/Modern International 现代国际
45/Hangzhou United Bank 杭州联合银行
49/Mongolia Wuhai Government Office Building 内蒙古乌海市政府性办公楼

R&D ON SCIENCE AND TECHNOLOGY ARCHITECTURE 科技研发建筑

53/Chengdu Longtan Headquarter Base 成都龙潭总部基地
57/Economy Zone of South Taihu Lake Headquarters 南太湖总部经济园区
61/Changxing South Taihu Lake Headquarter Office on No. A-01-03 Land in Economic Zone 长兴南太湖总部经济园A-01-03地块办公楼
65/Information Service Center of Shanghai Lingang Heavy Equipment Industrial Area 上海临港重装备产业区信息服务中心
69/Shanghai Academy of Automobile Engineering Stage-II 上海汽车工程院二期
73/Guiyang North Zhonghua Road Reconstruction Project 贵阳市中华北路城市复兴项目

目录

HOTEL ARCHITECTURE 酒店建筑

81/Front Coast Arts Hotel 前岸艺术酒店
85/Wuxi HILTON Yilin Hotel 无锡HILTON逸林酒店
87/Beijing Hongfu Hotel 北京鸿府宾馆
93/Ganzhou Gannan (Holiday) Hotel 赣州赣南（假日）酒店
97/Foshan Zongheng Hotel 佛山纵横大酒店投标
101/Sanya "Beautiful Crown" 三亚"美丽之冠"
103/Zigong Zhangjiatuo Totel 自贡张家沱酒店
105/Conceptual layout for phase three of Huangshan Songbai Golf Hotel 黄山松柏高尔夫酒店三期概念性规划

SHOPPING MALL 商业广场

109/Kunshan Boyue Piazza 昆山博悦广场
113/Wuyang Commercial Plaza 五洋商业广场方案设计
117/Commercial Architecture on Dongting Highway Block No.4 of Songjiang District of Shanghai 上海松江区车亭公路四号地块商业建筑
121/Suzhou Jing-Hang Canal Entertainment Center 苏州京杭运河娱乐中心
123/Shanghai Jiujiu Piazza 上海九久广场
127/Wuxi Yiju International Life Piazza 无锡逸居国际生活广场
131/Nanjing Wanda Piazza 南京万达广场
135/Wuxi Wanda Plaza 无锡万达广场
139/Tung Ying Building 东英大厦
141/Wuxi New Urban Business Piazza on Plot C4 无锡C4地块新城商业广场

Contents

MUSEUM, ART GALLERY AND EXHIBITION HALL 博物馆、艺术馆、展览馆

147/Design Scheme of Ai Weiwei's Works1　艾未未作品的设计方案
149/Franconia Jewish Museum　法兰克尼亚犹太人博物馆
153/MOCAPE-Shenzhen Modern Art Gallery and Urban Planning Exhibition Center　MOCAPE-深圳市当代艺术馆与城市规划展览馆
157/Guiyang Science & Technology Hall　贵阳科技馆（竞赛作品）
161/Art gallery of Sichuan Art Institute　四川美术学院美术馆
165/Langfang Grant Theater　廊坊大剧院
167/International Conference and Exhibition City　国际会展名城
171/Shenzhen University City International Conference Center　深圳大学城国际会议中心
177/Suzhou Conference Center　苏州会议中心
183/Layout of Hutang technology center　桂林漓东科技新城湖塘科技中心
187/New Embassy of the Republic of Turkey in Berlin, Germany　土耳其共和国驻德国柏林大使馆新馆
191/Guiyang Longdongbu International Airport terminal building　贵阳龙洞堡国际机场航站楼

HOSPITAL & CULTURE ARCHITECTURE 医疗文化建筑

197/Changzhou Xinbei People's Hospital　常州市新北人民医院
201/Hangzhou Xiasha Hospital　杭州市下沙医院
203/Liberation Army Jichang Road Branch of No.117 Hospital　解放军一一七医院机场路分院
207/Linchuan Cultural Garden　临川文化园
211/Guangzhou Children's Palace of Luogang District　广州市萝岗区少年宫
215/Huidong Old Cadre Activity Center　惠东县老干部活动中心
219/Science & Culture Center of Yunnan Chihong Xinzhe Incorporated Company　云南驰宏锌锗股份有限公司"科技文化中心"
223/Hefei Federation of Trade Unions Staff Activity Center　合肥总工会职工活动中心
227/New Teaching Building of College of Physics, University Rostock　罗斯托克大学物理学院新教学楼

目录

231/Foshan Complex of Public Culture 佛山公共文化综合体
235/Teacher's studio of Sichuan Fine Arts Institute 四川美术学院教师工作室

URBAN PLANNING 城市规划

243/Conceptual Planning of Binghu New Town, Dongtaihu Lake, Wuzhong, Suzhou City 苏州吴中东太湖滨湖新城概念规划
245/Mongolia Wuhai CBD Layout 内蒙古乌海CBD规划
249/Shenzhen Guanlan Print Base 深圳观澜版画基地
251/Qingdao Small Bay Super 5-Star Hotel Design and Coastline Layout 青岛小港湾超五星级酒店单体设计及海岸线规划
255/Urban Planning of Ecological Development and Regulation Section in Urban Section of Hutuo River, ShijiazhuangCity 石家庄市滹沱河市区段生态开发整治区城市设计
259/Xuhui Riverside Public Open Space 徐汇滨江公共开放空间
265/Hangzhou "Riverside City" Concept Plan 杭州"滨江新城"概念规划
269/Guigang Business Zone on No.GB-16 Land, Gangbei New District 贵港市港北新区GB-16地块办公商务区
271/Aachen Institute of Technology (Melaten Campus) Planning 亚琛工业大学校园(Melaten园区)规划

COMPREHENSIVE COMMUNITY 综合

279/Everbright International Center 光大国际中心
283/Shanghai Chunagzhan International Trade Center 上海创展国际商贸中心
285/Taikoohui Comprehensive Development, Tianhequ, Guangzhou286/广州天河太古汇综合发展项目
287/Shenzhen Line 4 (Phase 2) Property Developmen 深圳市轨道交通四号线二期沿线物业发展
289/Comprehensive Development of Plot 9, Danang, Vietnam 越南DANANG 9号地块综合开发
293/JingAn No.60 Lane 静安60#街坊
299/Nanyou Shopping Park, Nanshan District, Shenzhen City 深圳市南山区南油购物公园

OFFICE ARCHITECTURE

办公建筑

Hunan People's Radio Station Technology and Office Building

Location: Changsha city, Hunan province
Designed by: German S.I.C-Engineering Consulting Co., Ltd
Designer: Tim Bonnke
Frame: reinforced concrete
Land Area: 4 500 m²
Floor Area: 51 000 m²

The whole architecture includes a 23-storey circular main building and a 17-storey semi-circular annex building, connected by a three-storey multi-purpose studio hall. The hall on the first floor is designed for reception and external display. The technical-level functional rooms are arranged on the fifth floor of the main building, and the rest are office space with a clear delineation of functions of the various parts.

The inner courtyard design within the main building is able to create a good air circulation system and provide adequate natural lighting. The elevation uses ring-shaped long windows with the green color

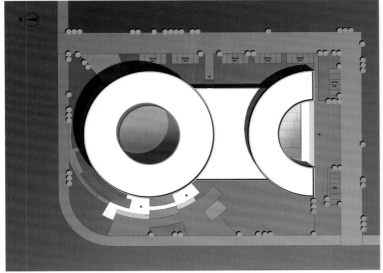
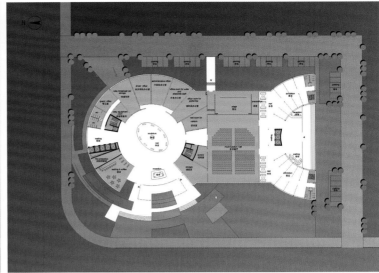

band as the main color of the wall stressing the building's features of environmental protection and energy conservation. The circular plane of each floor from north to south adopts concave convex changes in different degrees to form the visual center of the building and also create hanging gardens surrounding the building with the high and the low strewing at random. Looking at the building from the outside, one will feel a transparent and bright overall effect.

The design takes into account a number of energy-saving and environment-friendly programs to provide an excellent working environment to the employees in the office building. Within the scope of the construction base, in addition to the necessary entrance square and the parking lot, the design arranges a large amount of green space to optimize the overall environment. The exterior of the building uses insulating materials in exterior walls and windows with a good insulating property to reduce the inner energy dissipation. The exterior façade is equipped with special sun visor to reduce solar radiation and ensure indoor lighting. Inside the building the greening is mainly arranged in the scattered hanging gardens, in which the trees and vegetation is able to bring the building with adequate oxygen and reduce the internal temperature to provide a good leisure space for the staff. In summer, due to the higher direct-shot angle, the special opening structure of the hanging gardens can block a lot of direct solar radiation, and in winter with lower direct-shot angle, the sunshine will not be blocked by the opening and thus can enter the room in large amount.

The courtyard within the main building forms a good natural ventilation cycle; at the same time, the concrete pipe 10m deep underground continuously inhales the air through the engine; because of the constant temperature of the soil 10m deep underground, the air in the pipeline will be naturally cooled or heated by the soil here, and then transported in cycle to the output pipeline in the building (for every floors there are a number of cold air output ports) to meet the cooling or heating needs of all the rooms. Due to little gravity the existing hot air from the building will automatically rise from the courtyard to the top of the building and through the special opening at the roof escape. In addition, where possible, the building's outer surface (combined with the sun visor) or the roof can be installed with photovoltaic panels to generate electricity using solar energy to solve part of the electricity needs of the building.

湖南人民广播电台技术及办公大楼

项目地点：湖南省长沙市
建筑设计：德国S.I.C.-工程咨询责任有限公司
设计师：蒂姆·邦克 (Tim Bonnke)
结构形式：钢筋混凝土
基地面积：4 500 ㎡
总建筑面积：51 000 ㎡

整个建筑包括一栋23层的环状主楼以及一栋17层的半环状副楼，两栋楼通过三层高的多功能演播厅相连接。一层大厅有对外展示及接待功能。技术层功能性用房布置在主楼顶层5层，其余为办公用房，各部分功能划分清晰。

主楼的内天井设计能够形成良好的空气流通系统并提供充足的自然采光。立面采用环型的长窗，以绿色系色带为墙体主色调，强调了大楼环保和节能的特点。每层圆环平面的南北向处，采取了不同程度的凹凸变化，形成大楼的视觉中心，同时也营造出分布在建筑物四面高低错落的空中花园。从外部看建筑，会有通透、明亮的整体效果。

设计中考虑了多项节能及环保方案，能为在楼内办公的员工提供绝佳的工作环境。在建筑基地范围内，除了建筑入口广场和必要的停车场地外，设计布置了大量的绿化用地，优化整体环境。在建筑外部，采用隔热材料处理外墙，采用隔热性能良好的窗户，减少建筑内能量的散失。外立面安置特殊遮阳板，减少太阳辐射，保证室内采光。在建筑内部，绿化主要集中布置在高低不等的空中花园中，其间种植的树木和植被能为大楼带来充足的氧气，降低大楼内部的温度，为员工提供良好的休闲空间。夏季，由于太阳的直射角较高，空中花园的特殊开口结构能阻挡大量的太阳直射，而在冬季，太阳直射角较低，阳光不会被开口遮挡，能大量进入室内。

主楼内天井形成良好自然通风循环；与此同时，深埋在大楼下10m处的水泥管道通过引擎将地面空气不断吸入，根据地下10m处土壤常年恒温的特性，进入到管道中的空气将会被此处的土壤自然冷却或加热，而后循环输送到大楼内的输出管道（每三层设有多个冷气输出口），满足各个房间的制冷或供暖需求。由于比重较小，大楼内部已有的热空气则会自动由天井上升到大楼的顶部，并通过屋顶的特殊开口逸出。此外，在可能的情况下，大楼的外表面（可与遮阳板相结合）或屋顶可安装光电板，利用光能产生电能，以解决大楼的部分用电所需。

Shenzhen Changfu Jinmao Mansion Design

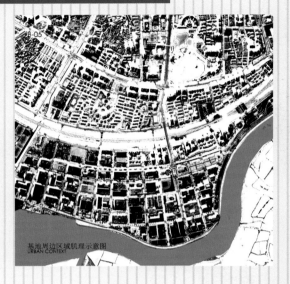

Location: Land B105-31, Futian Bonded Area, Shenzhen City
Designed by: French AUBE Architecture and Urban Planning Company
Land Area: 18.8 ha
Floor Area: 221 900 m²
Altitude: 350m

The project of Shenzhen Changfu Jinmao Mansion is located in the heart part of Bonded Area in Futian Shenzhen. It is adjacent to the north of Hong Kong spaced out by Shenzhen Bay, on the east of famous Mangrove Nature Reserve, at the west of the important Shenzhen Huanggang port. Here it goes into Yitian Road extending to the No. 3 entrance of Free Trade District, while on the north it directly connects to the Shenzhen Convention and Exhibition Center,

the Shenzhen Municipal Government and the central area of Futian. So the base is sitting on the south end of the city's principal axis with convenient traffic and prominent position.

The loop area on the south of Futian Bonded Area is just between Hong Kong and Shenzhen, which has got the intention of development from both Shenzhen and Hong Kong in mid & long-term planning. The blueprint is to build this waterfront treasure plot into a comprehensive mass of high-tech development and University City, which will have a tremendous growth in future foreground and create new opportunities for layout of Bonded Area, targeting orientation of projects here as well as the development mode.

It is designed to be an efficient, bran-new and human-oriented standing super high-rise building with full respect for the urban design in Bonded Area. Meanwhile it creates an urban space with the sense of region independence, combining the natural ecology and economy into one multi-functional integrated body. It shall be the site to enhance the urban culture loved by the public, at the same time it shall be the symbol of Shenzhen and Bonded Area in new century.

Basic design concept of being straight and slender will be taken here. It shall have a wide middle part with two small ends, which means it shrinks upward and downward based on the square plane. The

highest point of tower plane appears at the one-third dot of coping, which models the more dominant flowing up trends according to the golden section. Buildings almost covered by glass look rich in lights and colors, shining in the Peng City and spreading on. It seems like a plunging waterfall either standing far or near, bringing people the heroic sense that it is pouring down from thousands of meters high point.

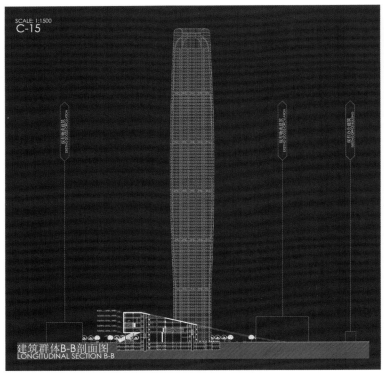

建筑群体B-B剖面图
LONGITUDINAL SECTION B-B

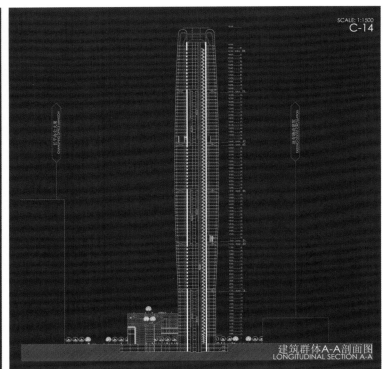

建筑群体A-A剖面图
LONGITUDINAL SECTION A-A

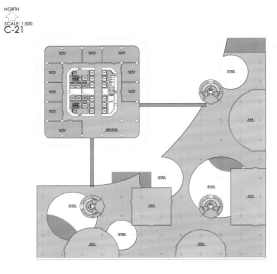

二层平面布局图
FLOOR-2 LAYOUT (LEVEL-12.00)

三层平面布局图
FLOOR-3 LAYOUT (LEVEL-18.00)

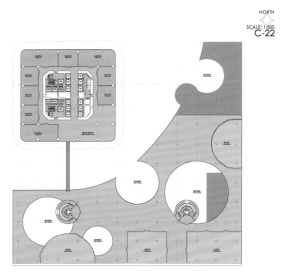

四层平面布局图
FLOOR-4 LAYOUT (LEVEL-24.00)

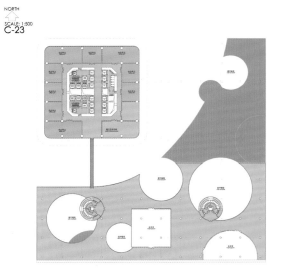

五层平面布局图
FLOOR-5 LAYOUT (LEVEL-30.00)

立面图

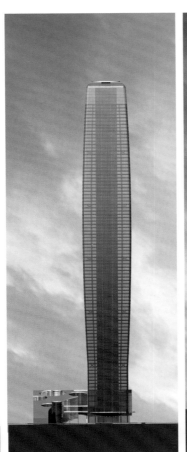
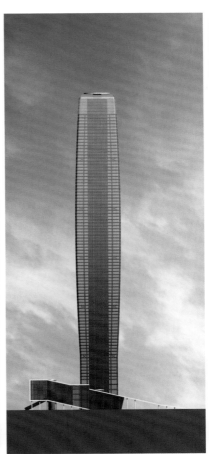
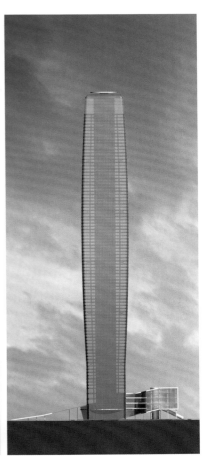

小透视
PARTIAL PERSPECTIVE

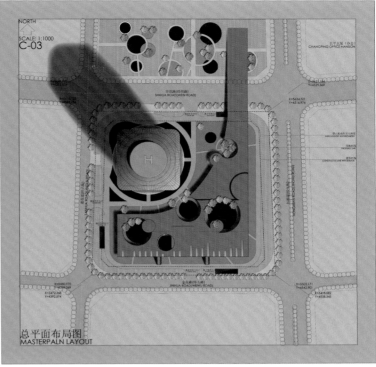

总平面布局图
MASTERPALN LAYOUT

深圳市长富金茂大厦建筑设计

位　　置：深圳市福田保税区B105-31地块
设计：法国欧博建筑与城市规划设计公司
用地面积：18.8ha
建筑面积：221 900 ㎡
高　　度：350m

　　深圳市长富金茂大厦项目位于深圳市福田保税区核心地段，南侧隔深圳湾与香港相邻，西侧为著名的红树林自然保护区，东侧为深圳重要的皇岗口岸，顺延保税区3号入口即是益田路，向北直接连接了深圳市会展中心和深圳市政府以及福田中心区，基地堪称位于城市主轴线的南尽端，交通便捷，位置显要。

　　福田保税区南侧的河套地区位于香港与深圳之间，在中远期规划当中，深港双方均有合作开发此区域的意向，规划蓝图是将这片濒水宝地建设成高新科技开发与大学城的综合体，未来前景甚为可观，这也同时给保税区的规划格局以及区内的项目定位与开发模式创造了新的机遇。

　　设计将拔地而起一座高效的、全新的、人性化的超高层建筑，充分尊重保税区的城市设计，同时创造具有独立领域感的城市空间，结合经济性与自然生态于一体的多功能综合体，它将是公众喜闻乐见同时又提升城市文化的场所，与此同时又是新世纪保税区以及深圳市的标志。

　　设计秉承修长挺拔的基本理念，两端小中间大，在方形平面的基础上，上下收分，在塔楼的顶部高度三分之一处为塔楼平面最大高度，依据黄金分割比塑造出更为高耸流动的上升动态，以玻璃为主的建筑群流光溢彩，光耀鹏城，恒远流传。远观与近赏，恰似一抹飞瀑倾空而泻，大有"飞流直下三千尺"之豪迈。

Shenzhen Konka R&D Mansion

Location: East of Southern District of Hi-Tech Zong, Shenzhen City
Designed by: French AUBE Architecture & Urban Planning Design Company
Land Area: 0.96ha
Floor Area: 80 000 m²
Floor Area Ratio: 8.3
Altitude: 100m

The Southern Area of the High-tech Zone where the project is located is land for urban construction at the east of Nanhai Avenue, south of Shennan Avenue with superior geographical condition, convenient transportation and wide range of radiation. The project site has many restrictive conditions and the land surface shows irregular trapezoid. The high floor area ratio, high threshold, irregular patches, as well as the uneven grade of the existing property management poses difficulty to the project's high-end positioning.

The design will create a highly efficient, new and user-friendly modern high-rise building with full respect for the urban design in the High-tech Zone while creating an urban space with a sense of

independent territory. The multi-functional complex combined with natural ecology and economy will become a place popular with the public and enhancing the urban culture, and at the same time will be the scenery and logo of the High-tech Zone and Shenzhen City in the new century.

In dealing with the relationship with the city, designers on the one hand respect the overall planning of the city; on the other hand actively create unique space identity and relatively independent internal public space for people entering the site to feel a strong sense of territory.

The design in conjunction with the urban space planning structure and characteristics of the surrounding building environment and natural environment uses cutting means on the basis of integrity to cut out different expressions in line with all the directions and functional demands. A large number of hanging gardens and shared spaces intersperse at the openings after each cut in the building body, creating people who work here a comfortable, easy-to-contact and nature-friendly space, which is also part of Konka's corporate culture.

Following the integration principle in architectural landscape design, a huge skin accompanied by clear and straight linear texture rises from the ground, becomes deformed, pulled and turns around to wrap up all the functional spaces while produce different sizes of shared spaces, and finally returns to the ground to form a seamless and coherent landscape system. The building is bred from the earth, and finally returns to the earth so that the boundary between the architecture and landscape is broken and the continuous dynamic space creates a stage for activities and events.

The façade on one side of Shennan Road is divided into two parts by the vertically distributed hanging gardens, forming an image of letter "K". This is a construction-based translation to Konka Group's logo, which forms a huge business card for the city and embodies Konka Group's brand image and design idea with advertising effect.

深圳康佳研发大厦

位　　置：深圳市高新区南区东片
设计单位：法国欧博建筑与城市规划设计公司
用地面积：0.96ha
建筑面积：80 000 ㎡
容积率：8.3
高　　度：100m

　　本项目所处的高新区南区为南海大道以东、深南大道以南的城市建设用地，地域条件优越，交通便捷，辐射范围广。项目地块限制条件较多，地块平面呈现不规则的梯形。高容积率、低限高、不规则的小型用地以及片区现有物业档次的参差不齐为项目的高端定位提供了难题。

　　设计将创造一座高效的、全新的、人性化的现代化高层建筑，充分尊重高新区的城市设计，同时创造具有独立领域感的城市空间，结合经济性与自然生态于一体的多功能综合体，它将是公众喜闻乐见同时又提升城市文化的场所，与此同时又是新世纪高新区以及深圳市的风景和标志。

　　在处理与城市的关系上，设计者一方面尊重城市的整体规划，另一方面积创造独特的空间识别性和相对独立的内部公共空间，使人们一进入地段就会产生强烈的领域感。

　　设计结合城市空间规划结构和周边建筑环境和自然环境特色，采用切割的手法，在整体性的基础上，剪裁出符合东南西北各个朝向和功能需求的不同表情。大量的空中花园和共享空间穿插在建筑体量中各个切割后的开口处，为在这里工作的人们创造舒适的、易于交往的、贴近自然生态的空间，这本身也是康佳企业文化的一部分。

　　遵循建筑景观的一体化设计原则，一张巨大的表皮伴随着挺拔清晰的线性肌理，从地面上升、变形、拉扯、回转，包裹了所有功能空间的同时，也生成大小不一的共享空间，最后又回到地面，形成景观系统，浑然一体，一气呵成。建筑从大地孕育而生，最后又回到大地，建筑和景观的界限被打破，连续动态的空间创造了活动和事件的舞台。

　　深南路一侧的立面被垂直分布的空中花园分割成若即若离的两部分，形成了一个字母"K"的意象，这是对康佳集团LOGO的一次建筑化转译，形成了一张巨大的城市名片，体现了康佳集团的品牌形象以及广告效应的设计理念。

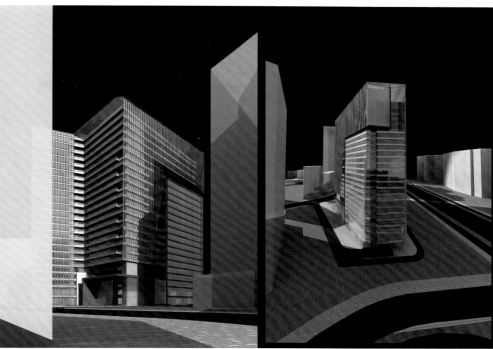

大堂透视 PERSPECTIVE　AUBE 欧博设计

Shenzhen Diamond Tower

Designed by French AUBE Architecture and Urban Planning Company
Land Area: 2.57 ha
Floor Area: 162 300 m²
Altitude: 300m

This is a challenging project.

Ground upwards to sky: starting form this viewpoint, design reflects two themes –ground and sky, their transition as well from different layers. For the vertical elevation, solid concrete is connected with the ground and lightweight material is concealed up to the sky; as to the landscape, the city's noise is shielded outside by the new urban park, static water scenery and bamboo forest.

Standing and stretching to sky: 318 meters' height for the new tower gives a new skyline to the city, which becomes the focus of the urban landscape. Now viewing from the almost built Shenzhen-Hong Kong Bridge, tower shows its forceful horizon outline in the daytime, concrete and glass material echoes the Nanshan Mountain and rosy clouds; while at night, elevation open to the bridge forms a light band pointing to the sky, galleries on the top floor and the restaurants are shining like jewelry, the whole building is attracting sight from the other side like a beacon, displaying the unique charm of seaside architecture.

Unique beauty: it respects for the overall urban planning for one aspect, while on the other hand it actively creates unique space identification and relatively independent internal public space in disposing the relation with city, which will provide people with a strong field depended sense on entering this region. It shall create a space of comfort, easy communication and closer to natural ecology for the people who work, reside or stay here.

The reason why it is called diamond tower is because it is not only the harmonious unity of heaven, earth, construction and human, but also a gathering place for wealth.

深圳钻石塔

设计：法国欧博建筑与城市规划设计公司
项目用地面积：2.57 ha
总建筑面积：162 300 ㎡
建筑高度：300 m

这是一个富有挑战性的项目。

由地及天：由此点出发，设计将从不同层面体现天、地两个主题及由地及天的过渡关系。在立面处理上，下部坚实的混凝土连接大地，顶层的轻盈材质隐入天空；在景观上，新的地景，即城市园林、静态的水景结合竹林将城市的喧嚣屏蔽于外。

顶天立地：318m的建筑高度使新塔楼塑造了新的城市天际线，并成为城市景观的焦点。从即将建成的深港大桥上观看：白天，塔楼显示出它挺拔的天际轮廓，混凝土和玻璃的材质呼应着南山和云霞，夜晚，面对大桥开敞的立面形成一道指向天空的光带，顶层的展廊与餐厅如同宝石般熠熠生辉，整个建筑如灯塔般吸引着对岸的目光，展示着海边建筑的独特魅力。

别有天地：在处理与城市的关系上，一方面尊重城市的整体规划，另一方面积极创造独特的空间识别性和相对独立的内部公共空间，使人们一进入地段就会产生强烈的领域感。为在这里工作，置业或暂住的人们创造舒适的，易于交往的，贴近自然生态的空间。

之所以命名为钻石塔，因为它是天、地、建筑与人的和谐统一，更是财富的聚集之地。

Shenzhen Kingkey Finance Center

Designed by: Huasen Architecture and Engineering Design Consultants Company Limited
Collaboratively designed by: TERRY & PARTNERS (UK) ARUPTE (UK)

Shenzhen Kingkey Finance Center mansion is totally 439 meters high with 97 layers, which is developed by Kingkey group. It is located in the bustling place of Luohu District, covering an area of 45,665 square meters and construction area of 553,493 square meters. It shall be a is a large building complex integrating Grade-A office, six-star luxury hotel, large senior commercial apartment and housing together. Upon its completion, it will become the tallest building in Shenzhen and also one of the world's tallest buildings at present.

Six-star hotel is located on the 75~97 layers of financial center's top part with building area of 35005 square meters. People can enjoy the exciting views of the city by the wide visual field, overlooking the litchi Park and watching the Shenzhen Bay directly as far as Hong Kong. Hotel lobby is set up on the 94th floor, the 93rd floor for Chinese restaurant, the 75th ~ 76th floors for chambers and multi-purpose halls, the77th ~ 90th floors for 250 guest rooms including 176 standard rooms, 44 executive suites, 28 panorama rooms and 1 president apartment.

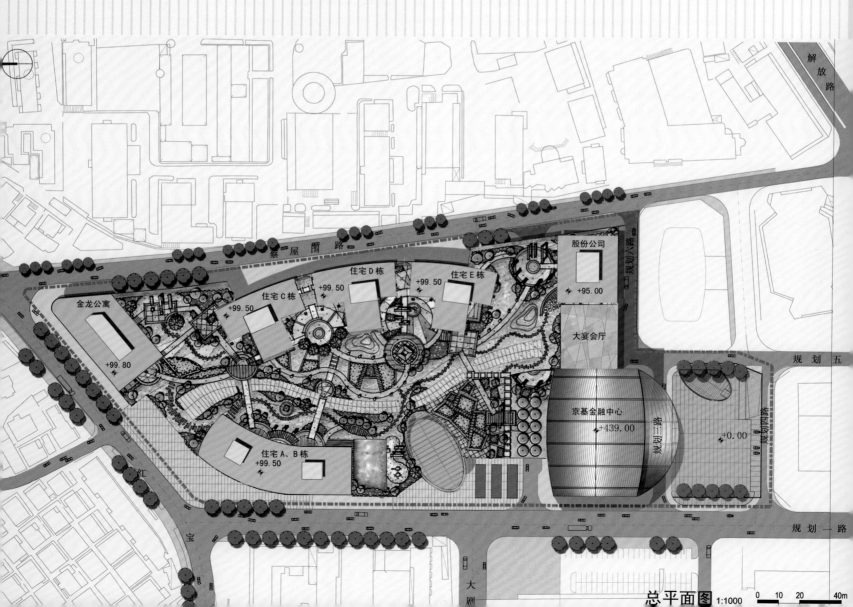

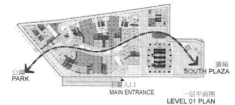

地下一层平面图　　　　　　　　　一层平面图　　　　　　　　　二层平面图
BASEMENT LEVEL 01 PLAN　　　　　LEVEL 01 PLAN　　　　　　　　LEVEL 02 PLAN

三层平面图　　　　　　　　　四层平面图　　　　　　　　　五层架空层平面图
LEVEL 03 PLAN　　　　　　　LEVEL 04 PLAN　　　　　　　LEVEL 05 PLAN

	商业中庭		设备用房		核心筒、疏散楼梯
	商业		车库、理货通道		卫生间
	库房		疏散通道		

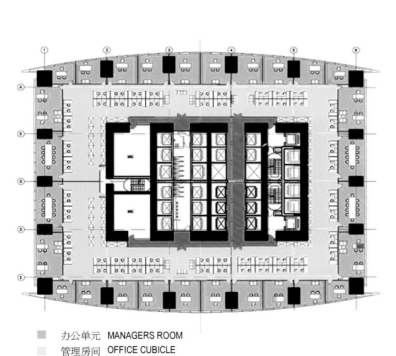

标准办公室剖面
TYPICAL OFFICE SECTION

拱肩镶板 SPANDREL PANEL
双层玻璃窗 DOUBLE GLAZED PANEL
双层地板 RAISED FLOOR
结构范围 STRUCTURAL ZONE
机电范围 SERVICE ZONE
吊顶 SUSPENDED CEILING
办公室净高 OFFICE CLEAR HEIGHT

办公单元 MANAGERS ROOM
管理房间 OFFICE CUBICLE
流通路线 OFFICE CIRCULATION

标准交易层剖面
TYPICAL TRADING FLOOR SECTION

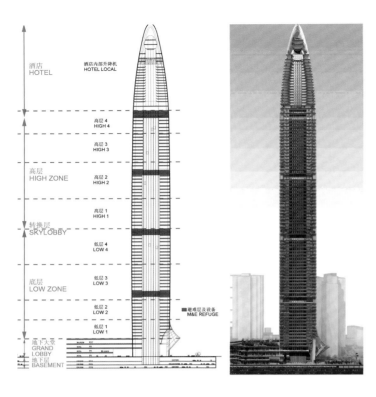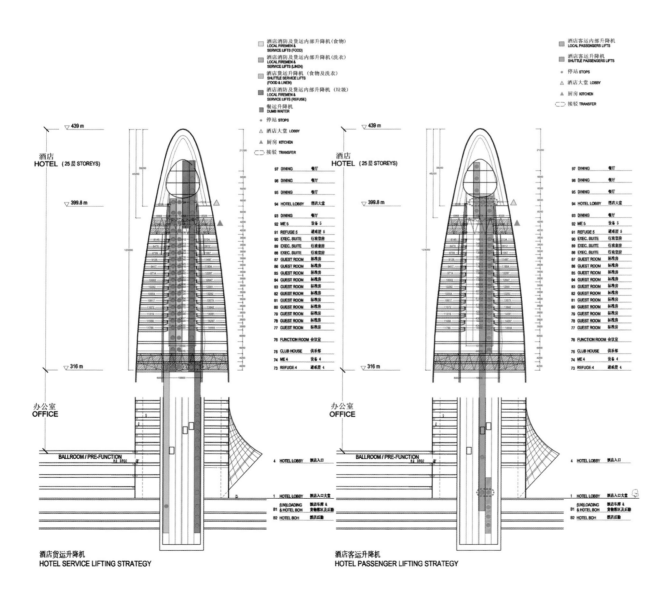

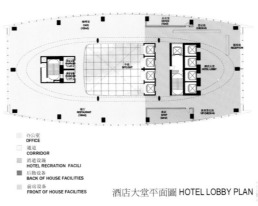

酒店大堂平面图 HOTEL LOBBY PLAN

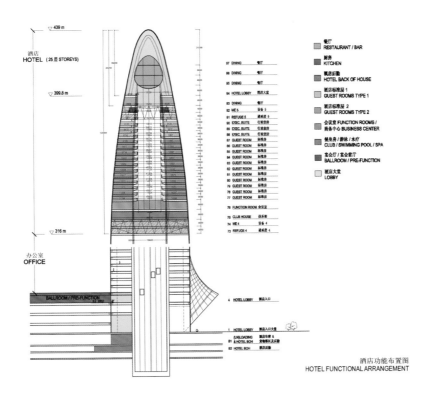

酒店功能布置图
HOTEL FUNCTIONAL ARRANGEMENT

深圳京基金融中心

建筑设计：华森建筑与工程设计顾问有限公司
合作设计：TERRY& PARTNERS（英）ARUPTE（英）

深圳京基金融中心大厦高439米，共97层，由京基集团开发，位于深圳市罗湖区繁华地带，占地45 665m²，总建筑面积为553 493m²，是集甲级写字楼、六星级豪华酒店、大型商业、高级公寓、住宅为一体的大型综合建筑群，建成后将成为深圳第一高楼，也是目前世界上最高的建筑之一。

六星级酒店位于金融中心顶部75~97层，建筑面积为35 005m²，视野开阔，可欣赏到令人心动的城市景观，俯视荔枝公园，远眺深圳湾直指香港。酒店大堂设在94层，中餐厅在93层，会所和多功能厅设在75~76层，在77~90层设有250套客房，包括标准间176间、行政套房44间、全景房28间和总统房1间。

Jinshi International Creativity Center

Location: Jinshi IT industrial zone of Dalian development zone
Planned and designed by: Shanghai Dblant Engineering Design and Consulting Co., Ltd
Land Area: 329 408m²
Floor Area: 407 706m²
Floor Area Ratio: 1.24
Building Density: 27.60%
Greening Ratio: 44.70%
Parking Space: 4 077

Jinshi International Creativity Center is located in the south of Jinshi IT industrial zone, with Jinshi Beach National Tourist and Holiday Resort Area famous as "Northeastern Jiangnan (South of the Yangtze River)" to the east. Abundant natural landscape resources are advantageous for the shaping of the environment of base.

With reference to the concept of fractal which is an artistic concept as the start of design, the design inserts comprehensive service cluster to each central functional area so as to break the functional structure of base into parts. Service cluster, as a shared communication platform, is able to exalt the convenience of each cluster.

The scattered comprehensive service clusters may be related together by a walking space in the middle to become a whole. This is the process of gathering parts together to form wholeness. Research cluster and innovation cluster are distributed in the northern and southern sides of the land.

The center are divided into four clusters as planned, which are research cluster, idea cluster, comprehensive service cluster and management cluster. Research cluster is in the north of the land being dedicated to introducing independent and large R&D organs and engineering centers, with low-rise and multi-storey research building as the main form. The cluster can offer each incoming enterprise an independent space. Idea cluster is placed to the south of the land. This cluster is trying to attract hi-tech enterprises and R&D institutes, as well as college technical zone and private incubators, in order to combine "industry", "university" and "research institute" into one. Comprehensive buildings with giant mass suitable for office, R&D, experiment and production purpose are the main forms, attached with business matching facilities. Comprehensive service cluster is placed in the middle of the land in the form of strips toward the south and north providing comprehensive service to the whole creativity center, including business conference, negotiation, catering, culture, leisure and commerce. It becomes the core of pedestrian flow. Management cluster is set in southwest and composed by houses for management and exhibition center.

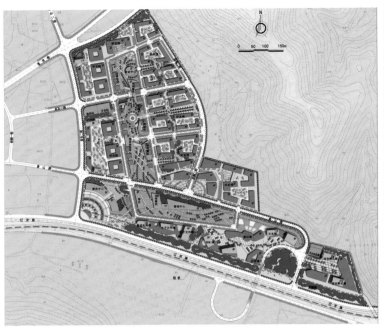

金石国际创新中心

项目地址：大连开发区金石IT产业园
规划/建筑设计：上海都林工程设计咨询有限公司
用地面积：329 408m²
建筑面积：407 706m²
容积率：1.24
建筑密度：27.60%
绿地率：44.70%
停车位：4 077

金石国际创新中心位于金石IT产业园的南部，基地以东是有"东北小江南"之称的金石滩国家旅游度假区，丰富的自然景观资源对基地的环境塑造非常有利。

规划设计参考艺术领域的分形概念作为设计的出发点，将综合服务总团穿插在各个功能区中间，使基地的功能组织化整为零。服务组团作为共享交流平台，提升每个组团的使用便利性。

分散的综合服务组团通过中间的步行空间相联系，使其自身也成为一个整体，这同时又是一个化零为整的过程。创研组团和创新组团分别布置在地块北侧和南侧。

金石国际创新中心规划分为四个组团，即：创研组团、创想组团、综合服务组团和会展管理组团。创研组团位于地块北部，着力引进独立的、规模较大的研发机构、工程中心等。建筑形态以低、多层研发楼为主，可使各入驻单位相对独立封闭。创想组团位于地块南部，着力引进高科技企业及研发机构入驻，也将引进大学科技园、民营孵化器等入驻，"产"、"学"、"研"相结合。建筑形态以体量较大的以及适于办公、研发、实验、生产的综合性建筑为主，并辅之以商业等配套设施。综合服务组团位于地块中部，呈南北向条状布置，为整个创新中心提供综合服务，如商务会议与洽谈、餐饮、文化休闲、商业等，其将成为小区域人流聚集的核心。会展管理组团位于地块西南部，由管理用房和会展中心组成。

Shenzhen New Mansion of China Sea Oil

Location: Shenzhen
Designed by: German S.I.C-Engineering Consulting Limited Liability Company
Designer: Shi Diwen
Land Area: 12 713 ㎡
Floor Area: 200 000 ㎡
Altitude: 200 m/160 m
Storey: 44 stories

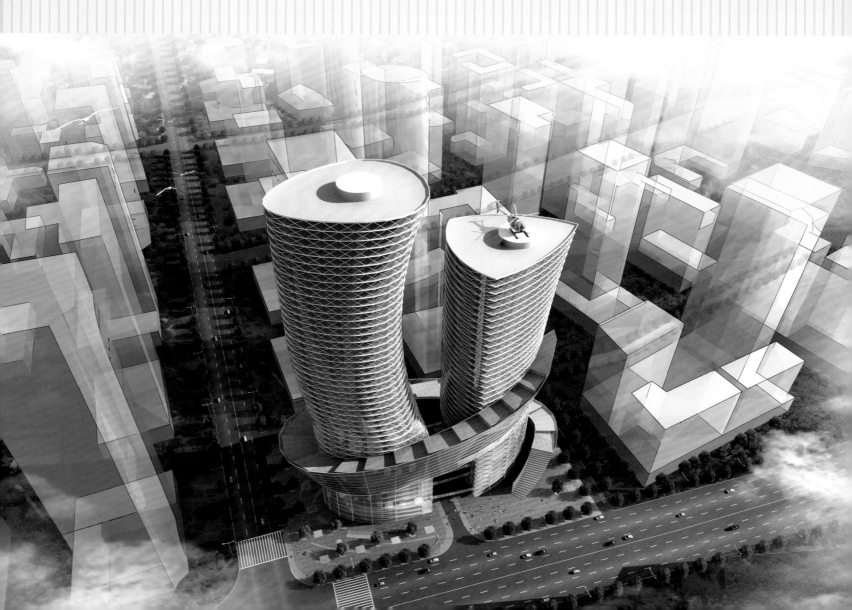

The land is embraced by the two torches, which is a symbol of the special bond between duration and natural energy. Through such a linkage, the surrounding landscape and architecture shall get closer to each other without any visual separation. There is a "climb up" trend for the podium compared to the opposite-oriented rotation in building's internal space. Three fourths of traffic volume is stretching into the underground space by passing shopping streets while stopping at Metro site.

Landscape continues spreading even entering into the building near the entrance of shopping street and Metro. The helical ribbons grounded part (or "roof" grounded part) is designed to be the sitting-out area that is regarded as a public space. The higher part of "roof" may be taken as a spreading of interior space, which is a private outdoor site. The rising spiral top part would be build to be the roof garden of the offices, which becomes a free space for relax. Because the roof is a changing open space, it may be used for a variety of activities.

Chief construction materials of outside elevation for the whole building are glass and steel, so are the spiral podium and the two towers. Use of natural materials such as wood and vegetation on the roof is soluble in one with the surrounding landscape, which can be reused in future.

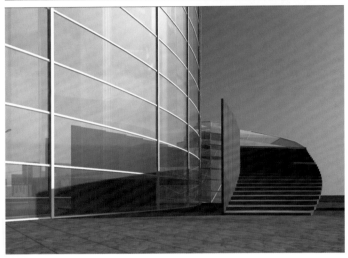

中国海油深圳新大厦

项目地址：深圳
建筑设计：德国S.I.C.-工程咨询责任有限公司
设计师：史迪文
占地面积：12 713㎡
建筑面积：约200 000㎡
建筑高度：200m/160m
建筑层数：44层

  两个火炬由大地拥抱，象征着可持续性与自然能源之间的特殊纽带。通过这样的联系，周围的景观与建筑之间的联系变得更亲近，在视觉上没有被分隔开来的效果。裙房部分带有"向上爬升"的趋势，而在建筑的内部的空间则有着反方向的旋转，交通⅔过商铺街往地下空间延续，止于地铁站点。

在靠近商铺街入口以及地铁站入口的地方，景观继续延续，并进入到建筑中。螺旋飘带接地的部分（或者说"屋顶"与地面相接处），可设置为供人休憩处，视为公共空间。在"屋顶"的较高处，可作为室内空间的延续，成为私有的室外场地。环绕上升的螺旋体顶部将作为办公室的屋顶花园，提供一个放松休憩的自由空间。由于其屋顶是一个不断变化的开放空间，因此可为多种使用空间提供场所。

建筑整体外墙立面使用的材料主要是玻璃和钢材，同时用于螺旋体裙房和两座塔楼。对于天然材料的使用，如木材和屋顶的植被，都与周围景观融为一体，并在将来可以继续开展再利用。

Eastern Blue Sea

Architecture Design: Shanghai Dulin Engineering Design and Consulting Co., Ltd
Dblant Desgin International

Scheme 1
Land Area: 32 940 m²
Floor Area: 114 900 m²
Floor Area Ratio: 3.49
Building Density: 35.40%
Greening Ratio: 36%
Parking Space: 546

Scheme 2
Land Area: 32 940 m²
Floor Area: 114847 m²
Floor Area Ratio: 3.487
Building Density: 33.85%
Greening Ratio: 47.53%

Parking Space: 600

Basic on the concept of "THE URBEN VILLAGE" and "HOPSCA" commercial development mode, the Eastern Blue Sea organically combines the main functions of the land with the existing functions around to make them complement each other and construct a unique advanced landmark business center with the integration of office, business, apartment and entertainment.

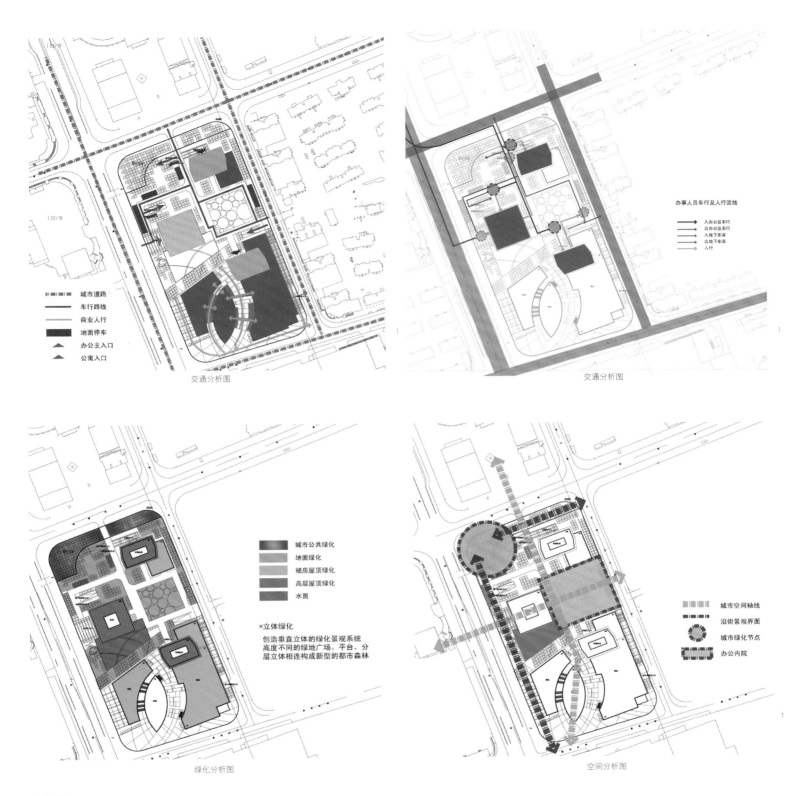

交通分析图   交通分析图

绿化分析图   空间分析图

东方蓝海

建筑设计：上海都林工程设计咨询有限公司

方案1
用地面积：32 940 ㎡
总建筑面积：114 900 ㎡
容积率：3.49
建筑密度：35.40%
绿地率：36%
停车位：546

方案2
用地面积：32 940 ㎡
建筑面积：11 4847㎡
容积率：3.487
建筑密度：33.85%
绿地率：47.53%
停车位：600

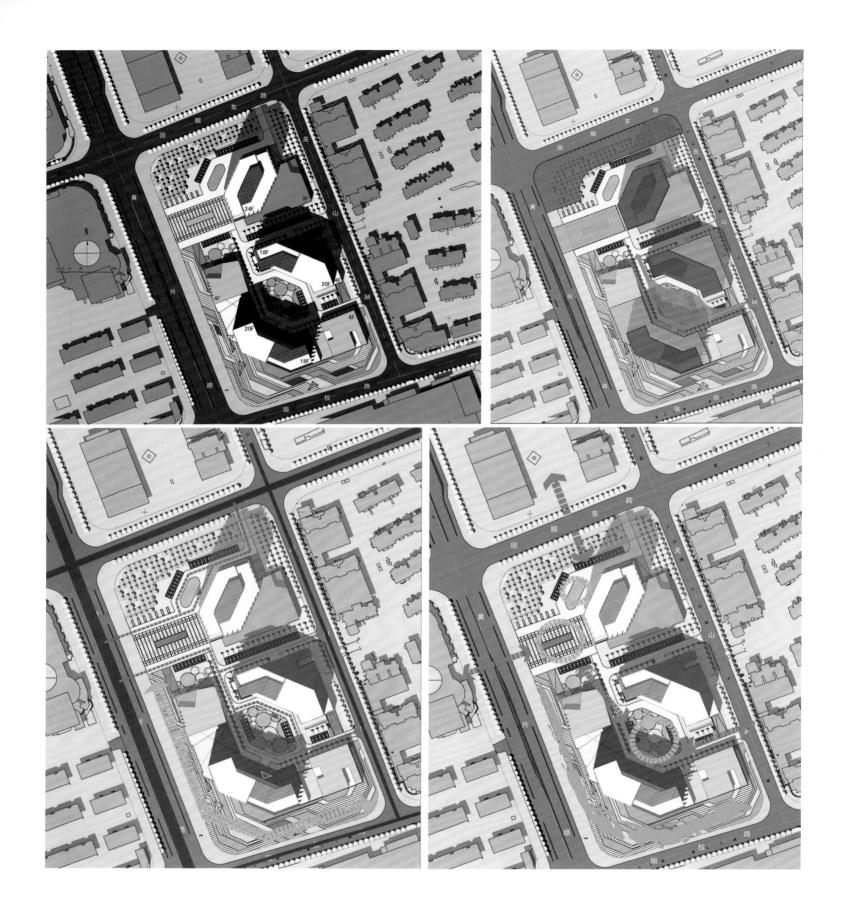

东方蓝海以"THE URBEN VILLAGE"为设计概念,集合"HOPSCA"商业开发模式,将地块主体功能与地块周边已存在的功能有机结合,互补不足,打造该地区独树一帜的集办公、商务、公寓与娱乐的地标性高级商务中心。

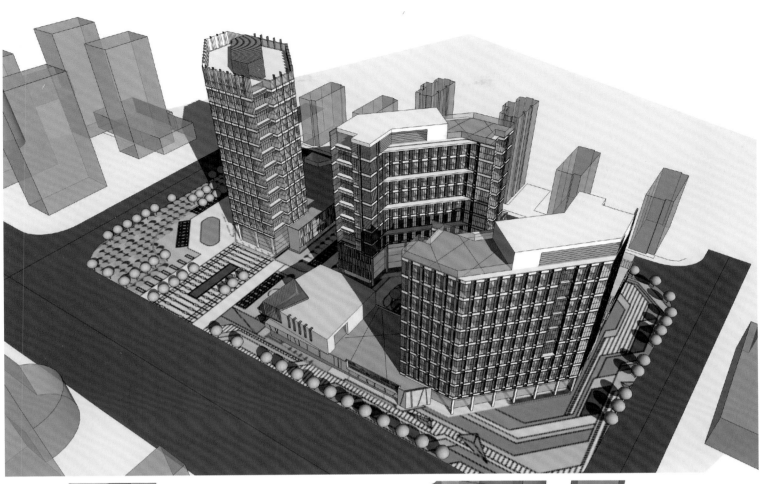
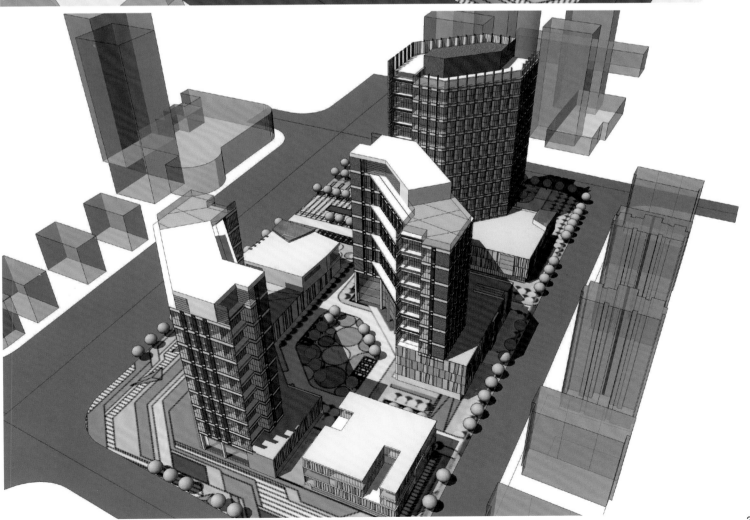

Binghai Zheshang Mansion

Location: Binghai New District, Tianjin City
Developer: Binghai Zheshang Investment Holding Co., Ltd
Architecture Design: ZPLUS Construction Layout and Design Company
Sight Design: Canadian Bansheng Designer Office

Land Area: 14 688 m²
Floor Area: 122 600 m²
Floor Area Ratio: 7
Residential Density: 40.49%
Greening Ratio: 30%
Parking Space: 618

After Shenzhen and Pudong, the third economic development wave of China is raised in the Binghai new district of Tianjin city and Xiangluo Bay is determined as the CBD of Binghai new district. With the acute business smell, Zheshang Company firstly builds up the Binghai Zheshang Mansion in the core area of Xiangluo Bay. The mansion is located in the business district of Xiangluo Bay, south of Hai River and east of Henan Road, in the Tanggu district of Tianjin city. It connects to Guihua Road to the east, Hengkang Road to the south, Yingbing Road to the west and Tuochang South Road to the north, occupying the core of Binghai CBD and the interchange of three major matching projects with convenient traffic and high business efficiency. Binghai Zheshang Mansion is composed by two towers of 26 stories and 30 stories which are connected by podium

buildings. Property form includes international hotel-style apartment, well-decorated studio, 5A-class office building and commercial street featured by international theme. The 3000 square meters' Zheshang chamber is set in the third floor of podium building including conference hall, Chinese and western restaurant, SPA and open garden.

Simple gray and pearl silver makes a straight and solid facade; steel and glass outline the upright sense with fine workmanship. Double and hollow super glazing brings wide view and provides efficient sound and heat insulation in environment-friendly way. Designers are to construct the modern classical building with the simplest style. The three elaborately-built luxurious halls differ from each other in style and serve as the purpose of business, living and chamber. Office building is equipped with a grand and solemn hall while the hall of hotel-style apartment pursues for the noble and elegant taste; hall of chamber is built with high-end casual atmosphere implying a steady-going and optimistic life attitude.

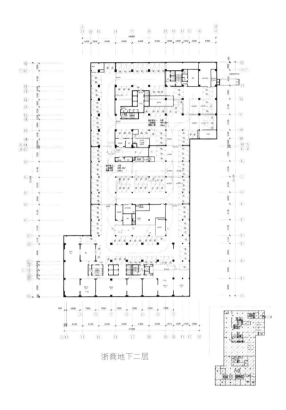
浙商地下二层

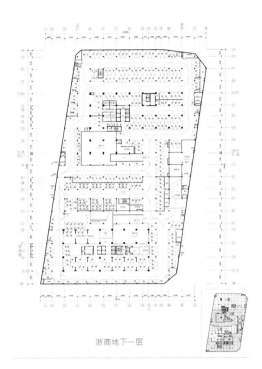
浙商地下一层

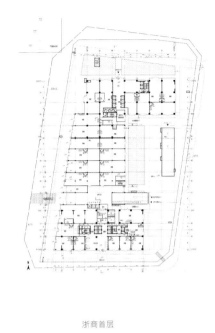
浙商首层

滨海浙商大厦

项目地址：天津市滨海新区
开发商：滨海浙商投资控股有限公司
建筑设计：ZPLUS普瑞思建筑规划设计公司
景观设计：加拿大班申设计师事务所
占地面积：14 688m²
建筑面积：122 600m²
容积率：7
住宅密度：40.49%
绿化率：30%
停车位：618

继深圳、浦东之后，中国第三次经济发展浪潮在天津滨海新区掀起，响螺湾被确定为滨海新区中心商务商业区。浙商凭借其敏锐的商业嗅觉，第一时间将滨海浙商大厦矗立在响螺湾的核心。滨海浙商大厦位于天津塘沽区海河以南、河南路以东的塘沽响螺湾商务区内。东面至规划路，南面至横康路，西面至迎宾大道，北面至坨场南道。雄踞滨海CBD核心，扼守三大配套工程交汇处，交通便捷，掌控商务效率。滨海浙商大厦由26层和30层到顶的两座塔楼构成，由裙楼相连。物业形态包括国际酒店公寓、精装工作室、5A写字楼、国际主题商街。3000m²的浙商会所位于滨海浙商大厦裙楼三层，会所内部设有会议厅、中西餐厅、SPA、透天花园。

滨海浙商大厦简约灰色与珍珠银建筑朗峻外立面，钢与玻璃勾勒挺拔质感，做工精细，双层中空超大玻璃窗，带来宽阔视野的同时，高效隔音隔热环保。以极简风格打造当代建筑经典。精心打造的三座奢华大堂，风格迥异，分别为商务、居住、会馆。写字楼大堂恢宏庄重，酒店式公寓大堂追求高尚优雅品位；会所大堂营造高端休闲氛围，淡定中笑看风云。

Modern International

Location: intersection of Jintian Road and Fuhua Road, Futian District, Shenzhen City
Construction design: America Kaipu Architectural Design Consultant (Shenzhen) Co., Ltd.
Land Area: 4 025.35m^2
Floor area: 67 548m^2
Floor area ratio: 14%
Greening Ratio: 74%
Parking space: 320
Altitude: 178.4m

This scheme is located in south of Shenzhen downtown. The Tower and annex match surrounding buildings to the fullest possible, and sotto portico, annex and main body of building designs respond to the whole layout in vertical line; axes of building and model of main body are stressed on mutual unity and harmony according to standard city layout of Shenzhen and guidance of SOM to city planning in this district; model of main body is concise and square, for one thing, it saves the most effective space for office, for another thing, the whole integrality is obvious and model is more straight. The northwest corner of the building is partially specially treated by triangle intersection, which impacts people's vision with diamond-like shine, while meeting the requirements of layout and concession and coordination of surrounding buildings. Unnecessary area is cut out in each standard storey to maximize utility rate. Height between floors is as high as 4.2m, spacious and not being restrained. The garden in the corner is fully made use of to share the plane as much as possible. Elevator passageway occupies 15% of the total area.

Sight design of this building consists of ground sight, annex roof garden, Tower roof garden and main body hanging garden. Ground sight, combined with sotto portico design, serves for business environment; annex roof garden forms a small internal and external connected world together with a 4.5-storey restaurant; Tower roof garden is special in architectural model and has broad eye shot. More designs have added in vertical metal-lined roof and special treatment of little transparent glass decoration. Firm stainless steel and aluminium alloy are used in façade of the building, and double-deck gray glass (treated by low-reflection filming) is used in solid glass curtain wall. Vertical line are predominant in building facade to achieve favorable economy and vision effect and match the building facade in the whole central area. On the surface of the building, size for windows is reduces for the purpose of economy, energy-saving and environmental protection. A series of high-efficiency glass, external blinds and intrinsic cycleventilation system are used for sun-shading facilities and heat energy conversion.

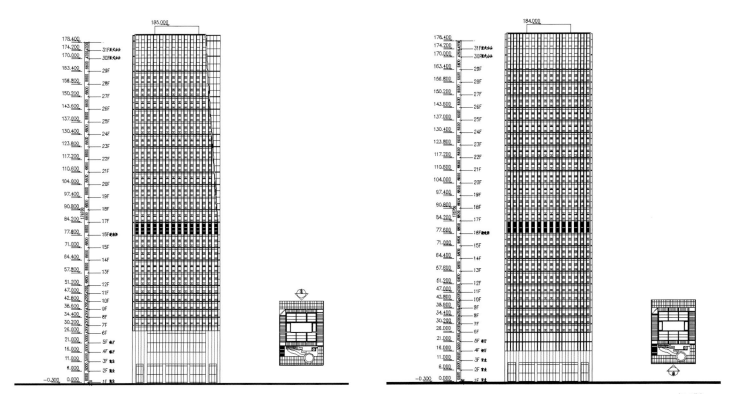

南立面图

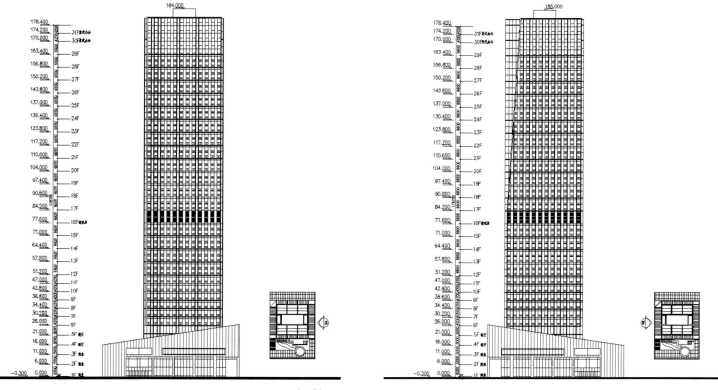

东立面图

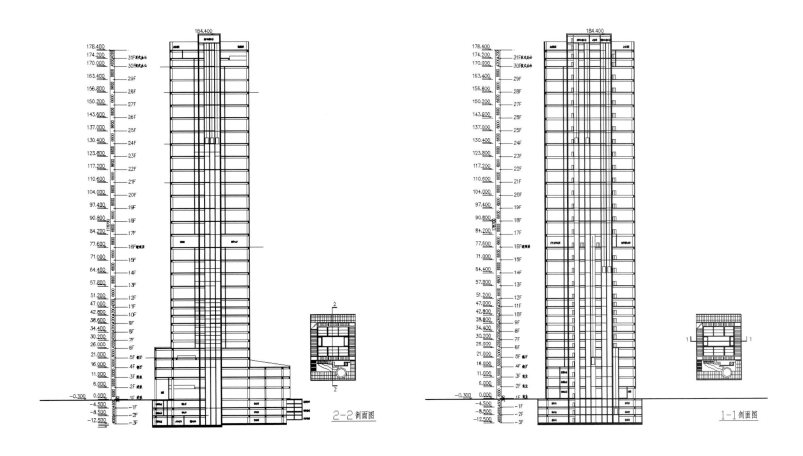
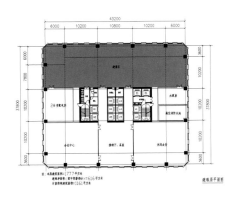
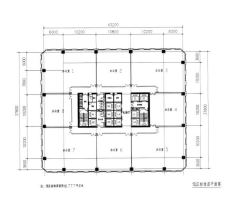
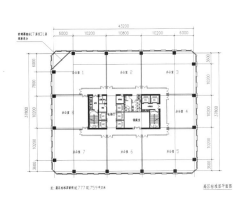
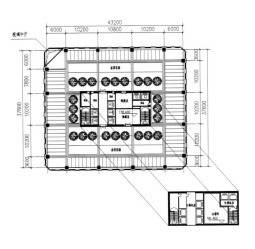
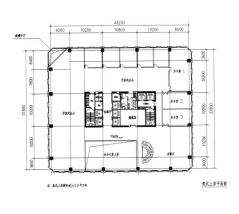
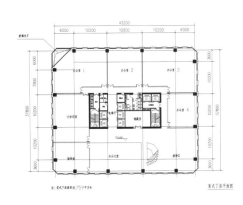

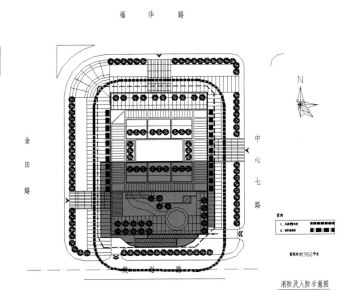
消防及人防示意图

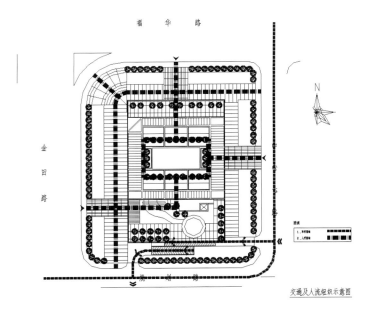
交通及人流组织示意图

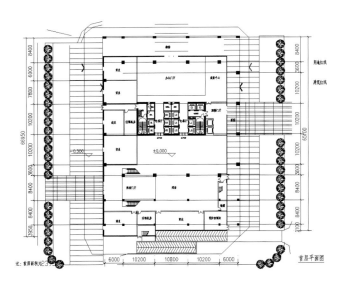
首层平面图

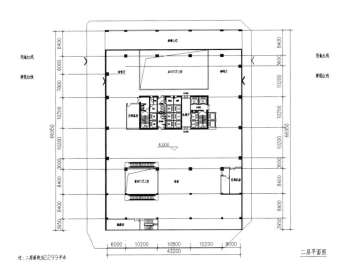
二层平面图

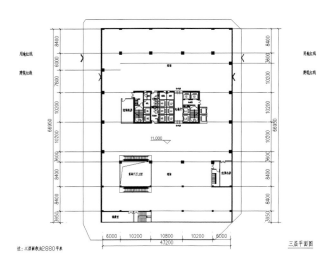
三层平面图

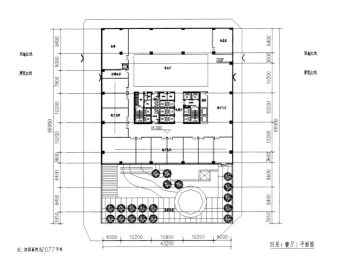
四层(餐厅)平面图

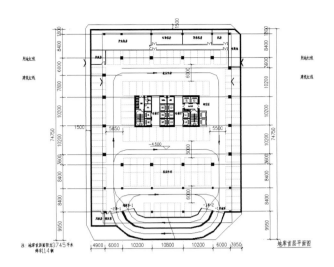
地库首层平面图

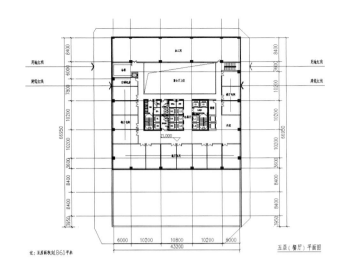
五层(餐厅)平面图

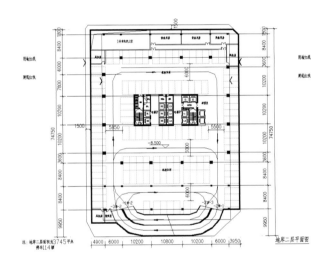
地库二层平面图

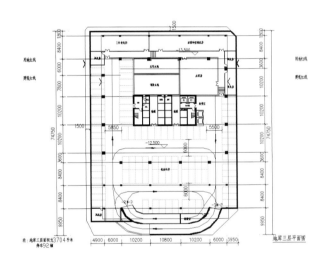
地库三层平面图

现代国际

项目地址：深圳市福田区金田路与福华路交汇处
建筑设计：美国开朴建筑设计顾问（深圳）有限公司
总用地面积：4 025.35m^2
总建筑面积：67 548m^2
容积率：14
绿地率：15%
建筑覆盖率：74%
停车位：320
建筑高度：178.4m

本方案地处深圳市中心区南区，塔楼与裙房位置尽最大的可能性协调周边的建筑，骑楼、裙房和建筑主体的竖向线条的设置均是呼应规划总体布局的要求；建筑的轴线和主体造型根据深圳市规整的城市布局结构和SOM对该区城市规划的指导，强调角部节点相互统一协调；建筑的主体造型简洁方正，一方面形成最有效的办公空间，另一方面使形体完整性更强，造型更挺拔，建筑主体西北角局部采用三角形切割的手法进行了特殊处理，在符合规划要求和与周边建筑退让协调的同时，给人以钻石般闪亮的视觉冲击。各标准层平面内消灭不必要面积，最大可能提高实用率。层高在4.2m，空间敞亮，不压抑。充分利用角部花园，尽可能多个平面共享。核心筒所占面积约占总面积15%。本建筑的景观设计分为地面景观、裙房屋顶花园、塔楼屋顶花园、主体空中花园四个部分。地面景观结合骑楼设计，主要烘托商业气氛；裙房的屋顶花园结合4.5层的餐厅形成内外连贯的小天地；塔楼屋顶花园具有独特的建筑造型元素和宽广的视野。设计更多的是竖向的金属线条搭配顶部和少量角部透明玻璃装饰的特殊处理，建筑立面采用竖挺不锈钢和铝合金饰面，立面玻璃幕墙采用双层灰色玻璃（均作低反射镀膜处理），建筑立面主要突出竖向线条，以达到了良好的经济性和视觉效果，和整个中心区建筑立面相统一。建筑表皮减少了开窗面积，节约造价，达到节能环保的目的。建筑的遮阳设施与热能转换可通过一系列高效能的玻璃、外部的遮阳百叶以及内循环通风系统来实现。

Hangzhou United Bank

Location: Middle Jiangguo Road, Shangcheng District, Hangzhou City
Designed by: No.1 Branch Company of Shenzhen General Institute of Architecture Design and Rearch
Designer: Zhen Fengzu, Li Xu
Land Area: 5763m^2
Floor Area: 36 462.52m^2
Floor Area Ratio: 4.96
Greening Ratio: 27.5%

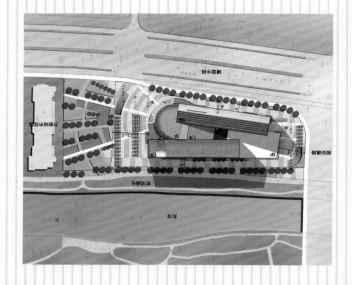

The Headquarter of Hangzhou United Bank is located in the west of Middle Jianguo Road, Shangcheng District of Hangzhou. The main tower building faces north in form of double-plate folding core tube, in which the northern plate building and the southern plate building form an alternate angle 8 ° to the northwest so that the main body opens the building's interface at the north west with the opening facing the north west, the direction to the city's leading landscape; meanwhile through changes in shape and the treatment of scattered-type hanging garden, the building falls in dialogue relations with urban spatial pattern, which totally changes the pattern of the side façade of plate building facing urban trunk road, forming an unusual architectural style.

The standard layer adopts plate in-tube layout with both ends open to form sound ventilation and lighting and made into scattered-type hanging garden. The elevation style seeks to be simple with no lack of beauty; the dual-plate high-level tower through the use of end-cap at the corner to produce different effect. The treatment of terrace hanging garden enriches the level of the building and also strengthens the tower's straight and tough image; the connection of tower and podium highlights the interspersing relations between masses in concise and generous manner to fully publicize it as the architectural character of the city portal.

The building design aims to create a multi-layered green landscape space and viewing space. Through podium roof greening and the treatment of green space on the tower roof viewing floor, the design creates an all-directional viewing space and landscape space to strengthen the communication between internal and external space, thereby creating a unique and personalized interactive shared space to further embody the positioning and objective of viewing architecture and landscape architecture design.

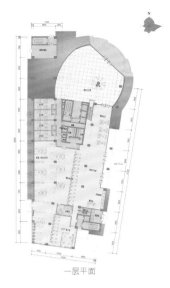
一层平面

二层平面

三层平面

四层平面

北立面

西立面

南立面

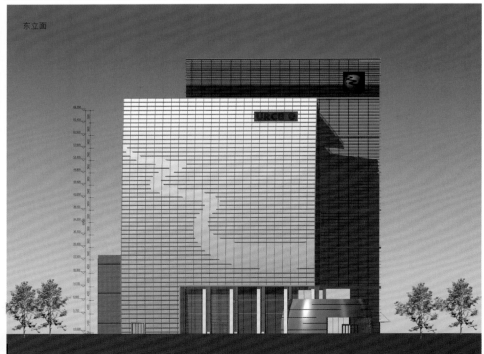
东立面

杭州联合银行

项目地址:杭州市上城区建国中路
建筑设计:深圳市建筑设计研究总院有限公司第一分公司
设计师：郑丰足、李旭
用地面积:5763m²
建筑面积:36 462.52m²
容积率：4.96
绿化率：27.5%

杭州联合银行总部大楼位于杭州市上城区建国中路以西,建筑中主楼塔楼坐南朝北,呈双板式夹核心筒布局形式,其中北面板式楼与南面板式楼向西北呈8°错角,使主体在西北面打开了建筑界面,开口朝西北面城市主导景观方向,同时通过造型的变化,错落式空中平台花园的处理,这样与城市空间形态产生对话,完全改变了板式建筑侧立面对城市主干道的格局,形成了非同一般的建筑造型。

标准层采用板式中筒式布局方式,板式的两端打开形成良好的通风和采光,并做成错式的空中花园。立面造型追求简洁,但是不乏精美,双板式高层塔楼通过转角收边的运用,产生不一样造型效果。采用叠层式空中花园的处理,丰富了建筑的层次,也强化了塔楼坚韧挺拔的建筑形象,裙楼与塔楼的交接突出体块间的穿插关系,手法简练大方充分张扬其作为城市门户的建筑个性。

建筑以营造多层次的绿化景观空间和观景空间为设计目标。通过对裙楼屋面绿化及塔楼屋顶观光层绿化空间的处理塑造全方位的观景空间和景观空间,加强了建筑内部与外部空间的沟通,从而形成独具特色和个性化的互动共享空间,进一步体现观景建筑和景观建筑的设计的定位和目标。

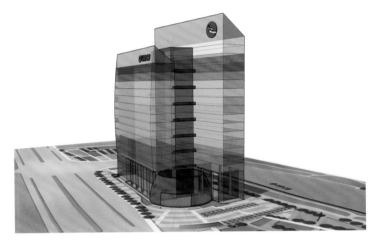

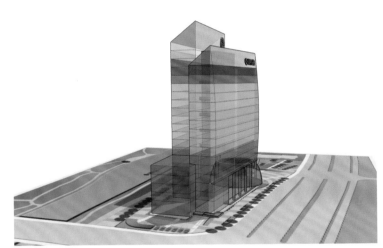

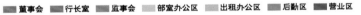

Mongolia Wuhai Government Office Building

Location: Wuhai city, Mongolia
Designed by: German S.I.C-Engineering and Consulting Limited Liabilities Company
Designer: Stephan Klabbers, An Jijun, Chen Jiayu
Land Area: 8 690 m²
Floor Area: 23 400 m²
Floor Area Ratio: 2.7

The plane layout looks L type with two construction masses inserted into each other to enrich the street-facing facade of construction. Passageway is set at external side to facilitate the in and out workers; the street-facing facade is decorated with aluminum plastic board to make it appear clean and generous giving prominence to office atmosphere; a yard is naturally formed inside, with restaurants designed to be hemicycle in order to increase the enclosing sense of space; the design is trying to build a warm space inside to offer a comfortable environment to workers.

To adapt to the cold weather in the north and save energy, we install a solar panel on the roof to make full use of the clean and natural energy; each storey is installed with auto-sensitizing and steel sun-breakers and double glazing to avoid direct sunlight.

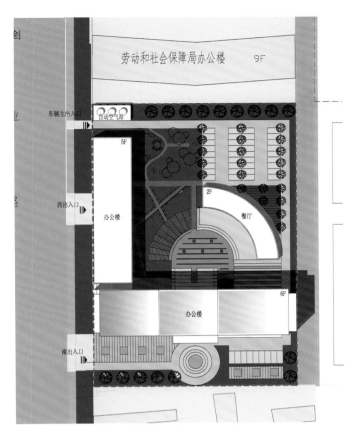

内蒙古乌海市政府性办公楼

项目地点：内蒙古乌海市
设计：德国S.I.C.-工程咨询责任有限公司
设计师：史迪文（Stephan Klabbers）、安继君、陈佳瑜
总用地面积：8 690 m²
总建筑面积：23 400 m²
容积率：2.7

项目平面布局呈L型，两个体块相互穿插，丰富沿街建筑的立面。外侧设置出入口方便工作人员出入，沿街外墙采用铝塑板，使之显得干净大方，突出办公氛围；内部自然围合成一个庭院，同时将餐厅设计为半圆形状，增加空间围合感，力争在内部营造一个温馨的空间，给工作人员一个舒适的办公环境。

为了适应北方寒冷气候，减小能耗，在屋顶加太阳能板，充分利用无污染的天然能源；每层加自动感光百叶钢结构遮阳板，避免夏季阳光直射以及采用双层玻璃。

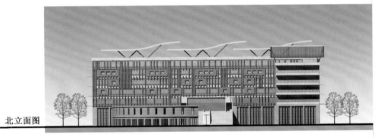
北立面图

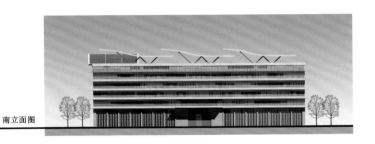
南立面图

东立面图

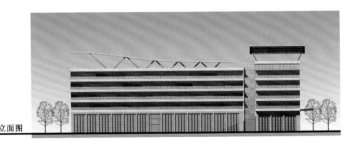
西立面图

R & D ON SCIENCE AND TECHNOLOGY ARCHITECTURE

科技研发建筑

Chengdu Longtan Headquarter Base

Location: east of Chenghua district, Chengdu city, Sichuan province
Developer: Chengdu Longtan Yudu Industries Co., Ltd
Planned and Designed by: Beijing A&S International Architecture Design and Consulting Co., Ltd
Designer: Yuli, Zhangbing and Renling
Land Area: 2 021 542 m^2
Floor Area: 3 438 164 m^2
Floor Area Ratio: 1.45
Building Density: 24.69%
Greening Ratio: 35%
Building Altitude: 200 meters

Chengdu Longtan Headquarter Base is built in the east of Chenghua district, occupying 1.8 million square meters. The building mainly serves for the office of headquarter together with hotel, business, apartment-style office, hotel-style apartment and other supporting functions. According to the analysis on the situation of the land and surrounding area and the economic characteristics of headquarter, designers have confirmed the direction of the new urban syntheses. It shall be ecology-friendly, multi-center, diverse, comprehensive and economy-led. There are three major landscape belts in the land naturally leading the scenery resources into the headquarter base; meanwhile, the designers apply and develop current water resource to form a natural and ecological layout along the water, and combine expert building with natural scene to make full use of the ecological environment inside the base. Secondly, basic on the analysis on urban traffic and functions, three major urban central areas are naturally formed. Land 1 naturally becomes the demonstration area of the headquarter base and land 3 is divided into two central areas by the junction of the three main programmed roads. For the existence of underground line in the northern side, traffic condition and function become more important. An ecological urban center will be formed in the south for the presence of landscape axes and the park being planned.

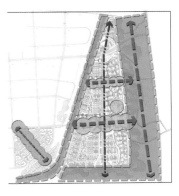
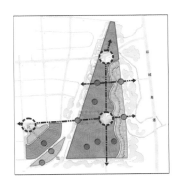

成都龙潭总部基地

项目地址：四川省成都市成华区东侧
开发商：成都龙潭裕都实业有限公司
规划设计：北京翰时（A&S）国际建筑设计咨询有限公司
设计师：余立、张兵、任凌
占地面积：2 021 542 m²
建筑面积：3 438 164 m²
容积率：1.45
建筑密度：24.69%
绿地率：35%
建筑高度：200m

成都龙潭总部基地位于成华区东侧，占地180万m²，功能以总部办公为主，涵盖酒店、商业、公寓式办公、酒店式公寓及其他配套功能。通过对用地及周边现状的分析，并结合现今总部经济的发展特点，确定了其生态性、多中心、多元化、综合性、以总部经济为主导的城市新综合体的规划方向。用地内形成三条主要的景观带，将景观资源自然地引入到总部基地内部，同时利用并开发现状的水资源，沿水系形成自然的原生态的规划形势，将独栋专家楼与自然景观结合，将总部基地内的生态性最大化。其次通过对城市交通和功能的分析，自然地形成三个主要城市中心区，一号地自然地成为了总部基地的示范区，三号地沿三条主要规划路的交叉点形成两个中心区；北侧由于地铁线的存在，更注重交通性和功能性；南侧由于景观轴线的存在和规划公园的因素，形成一个生态型的城市中心。

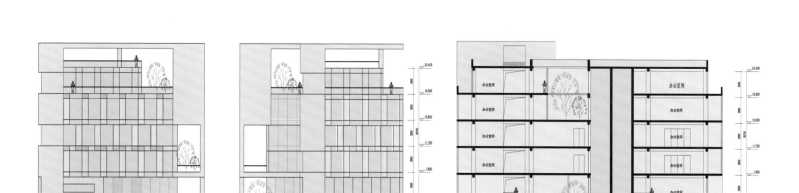

立面图 剖面图

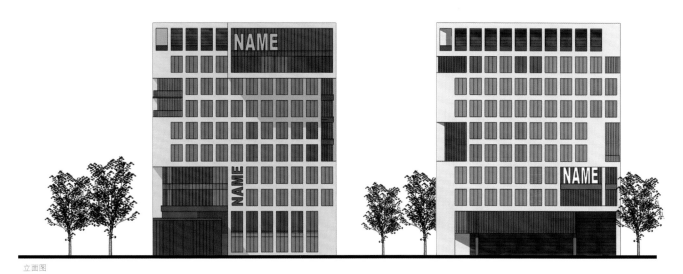

立面图

Economy Zone of South Taihu Lake Headquarters

Planned / Designed by: Dblant Desgin International

The provided land amount in old town area is very limited due to land restrictions in urban section of Changxing presently. Commercial or municipal reconstruction is taken as the chief item, main development direction of which includes new district in the north and headquarters of the economic zone in the east. As line side of Mingzhu Road opened in the future, it has significant superiority either from its market viewing point or the launching amount and competition.

As the investment intensity of Changxing County increases further and headquarters of the park gradually forms, development prospects are very promising. South Lake Taihu Headquarters Economy Zone takes "simple, modern" as the design concept for architectural style to create a strong atmosphere of economic zone. Design of landscape goes with the construction's simple, modern style based on spiffy lines. There is also an appropriate amount of hard landscape and vegetation, which sets off the architectural style and further enhances the land value in ecological and environmental quality at the same time.

南太湖总部经济园区

规划/建筑设计：都林国际设计

目前长兴由于市区用地等方面的限制，老城区供应的土地量相当有限，并以商业或市政改造项目为主，而发展方向重点为北面的新区和东面的总部经济园区，作为未来开放重点的明珠路沿线，无论从市场看点还是市场投放量、竞争度来讲都有着显著的优势。

随着长兴县招商引资力度的进一步加大，总部园区的逐渐形成，开发前景非常看好。南太湖总部经济园区建筑风格以"简洁、现代"为设计理念，营造强烈的总部经济园区氛围。景观设计配合建筑的简洁、现代风格，以明快利落的线条为主，结合适量的硬质景观和植被，在衬托建筑风格的同时，在生态及优质环境方面进一步提升地块价值。

Changxing South Taihu Lake Headquarter Office on No. A-01-03 Land in Economic Zone

Architecture Design: Dblant Desgin International
Land Area: 6 099 m^2
Floor Area: 22 747.7 m^2
Floor Area Ratio: 3.2
Building Density: 35.4%
Greening Ratio: 23.8%
Parking Space: 135

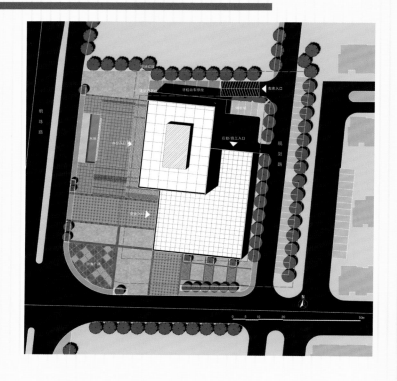

The economic zone No.A-01-03 land of Changxing Taipu Lake Headquarter is located in the economic zone with Beixi Road to the north and Mingzhu Road to the west. The place is with convenient traffic and advantaged predominance. The product types of this project are mainly for business and office purpose. Parking lot is set underground; podium building is a wide business concept, including hotel, office hall, catering, entertainment and conference center; tower accommodates apartment-style hotel and offices, so that the project is able to serve for multiple business purposes such

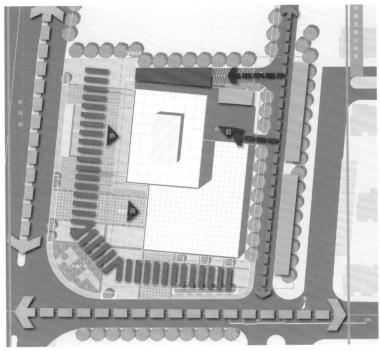
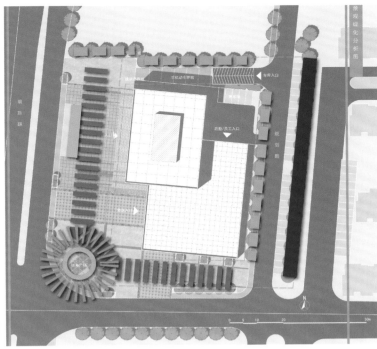

as office, lodging, negotiation, investment-attracting and conference. Meanwhile, the future property right and management have also been taken into consideration; multiple functions can meet different demands; comparatively centralized in the area with matching services, they have bright properties.

Basic principles of planning:
1) Reasonably engage with urban planning system to make full use of geographical and environmental resources;
2) Give consideration to the rhythm of horizontal outline;
3) Stress on the concept of landscape ecology basic on the characteristics of land and attach importance to the sustainable development of working environment and ecological environment.

Style of Architecture

Basic on its functions, being simple and modern is treated as the design concept to construct a great atmosphere in the economic zone. The facades of tower and annex are treated with uniform and integral design and modern and compact geometric combinations are adopted to boldly create the landmark outline of façade. Glass and stone are the main materials and make the whole architecture transmit a kind of steady-going, high-class, modern and fashionable temperament. In design, function of each section is reasonably organized to form a comprehensive building; in plane, enclosure layout is adopted to maximize the economic benefit of frontage and meanwhile create a complete functional space inside. The relation between the building and street is emphasized and dark-gray pattern is applied to organize the language of façade. Building design and internal design are unified.

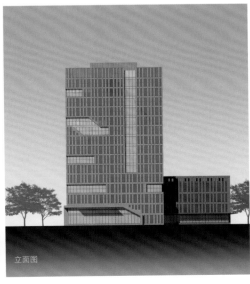

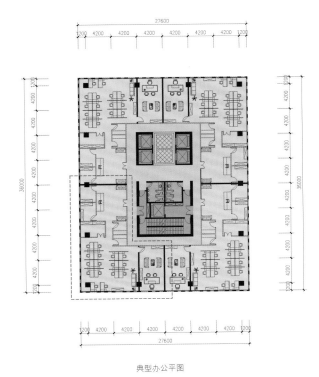

典型办公平图

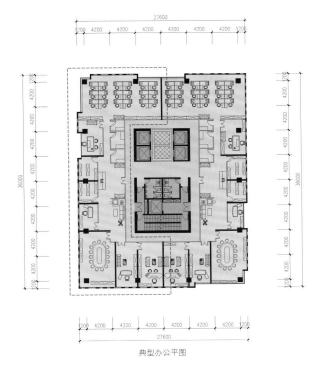

典型办公平图

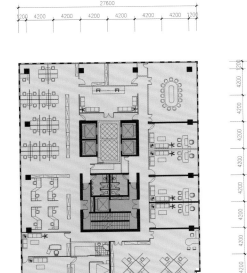

典型办公平图

长兴南太湖总部经济园A-01-03地块办公楼

建筑设计：都林国际设计
用地面积：6 099 m²
总建筑面积：22 747.7m²
容积率：3.2
建筑密度：35.4%
绿化率：23.8%
停车位：135

长兴南太湖总部经济园A-01-03地块位于长兴南太湖总部经济园区内，北临北溪大道，西临明珠路，交通便利，具有得天独厚的优势。项目产品类型以商业和办公为主。地下设置停车库；裙房为大商业概念，包含宾馆和办公大堂、餐饮、娱乐以及会议中心；塔楼部分包含公寓式酒店以及办公，这样可综合满足办公、住宿、洽谈、招商、会议等多方面的商务需求。同时考虑到今后物业产权和管理，多样化的功能满足不同的需求，相对集中在城市服务配套领域，具有鲜明的属性。

规划布局基本原则：
1）与城市规划体系合理衔接，充分利用地理和环境资源；
2）设计中考虑天际轮廓线的节奏与韵律；
3）结合用地特点，强调景观生态学的概念，注重办公环境和生态环境的可持续发展。

建筑风格：

在满足功能的前提下，建筑风格以"简洁、现代"为设计理念，营造强烈的总部经济园区氛围。

对塔楼及裙房的外立面进行统一整体的设计，运用现代简洁的几何组合，大胆做出能产生地标效果的外立面造型。材料以玻璃和石材为主，使建筑整体透射出既稳重高档又现代时尚的气质。在设计上通过合理组织各部分的功能布局，形成一个综合性的建筑体，平面上采用围合的布局，争取到最大临街面的经济效益，同时形成内部完整的功能空间。强调建筑与街道的关系，采用深灰色格调组织立面语言。建筑设计，室内设计相统一。

Information Service Center of Shanghai Lingang Heavy Equipment Industrial Area

Location: Lingang Heavy Equipment Industrial Area
Designed by: Dblant Desgin International
Land Area: 4 883m²
Floor Area: 12 299 m²
Floor Area on the ground: 9 757 m²
Floor Area under the ground: 2 542 m²
Floor Area Ratio: 2.0
Building Density: 9.8%
Greening Ratio: 30.2%

The Information Service Center of Shanghai Lingang Heavy Equipment Industrial Area has five floors with Huangsha Harbor in the west. It is a versatile comprehensive office building mainly used for communication and information computer room, communication service and business office.

The design concept is described as follows: several rectangular blocks with similar shapes are pieced together to create a square, from which a block is arbitrarily taken to place on the top of the whole square on the premise that the whole square will not collapse, and it continues by analogy. This concept is derived from the consideration about communication era. Things of modular form

may exist in various aspects of the communication industry. Through modular superposition, people's work will be more effectively simplified and a more relaxed environment will be available. Meanwhile, the potential pitfalls brought about by the modular things shall be on the alert, for any segment which goes wrong will lead to severe consequence to the overall situation.

The project is taken as the first building developed within the plot, which requires distinctive character as well as harmony with the original buildings in the south of the plot.

The layout of adytum enclosure is adopted for the building, the inner functions of which are arranged around the adytum to achieve good day lighting and ventilation and to effectively strengthen the vitality of the entire building. The air garden enlarges the space over the adytum of the building to avoid narrowness, which provides people with a more spacious and comfortable feeling from the space perception.

Three hallways are designed on the first floor, namely consuming business hallway, office hallway and grocery hallway, which distinguishes different steams of people and commodity streamlines. Areas are relatively dependent from each other and associate with each other through vertical traffic.

The technique of unevenness contrast as well as space-substance juxtaposition is employed in the design to shape the body. The building is mainly based upon substance, while the volume of space primarily reflects at the exit and the entrance of the building. Due to the location of the building, exquisite design of its roof (the 5th facade of the building) is implemented. The design concept of the roof and the entire architectural form are merged into a single whole.

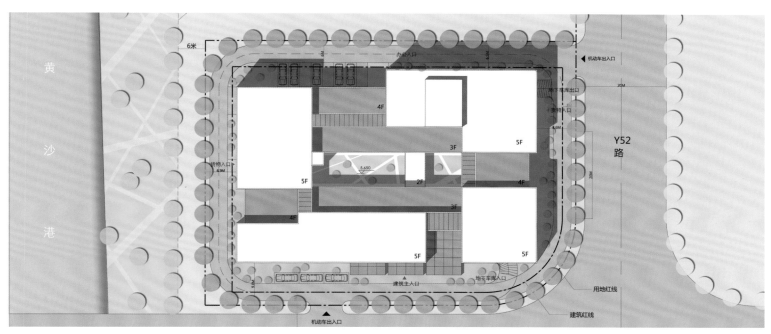

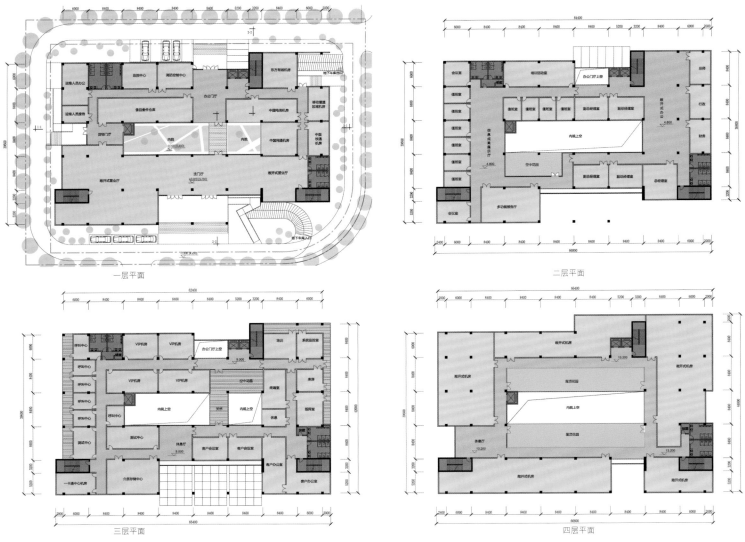

一层平面　二层平面　三层平面　四层平面

上海临港重装备产业区信息服务中心

项目地址：临港新城重装备产业区
建筑设计：都林国际设计
总用地面积：4 883m²
总建筑面积：12 299m²
地上建筑面积：9 757m²
地下建筑面积：2 542m²
容积率：2.0
建筑密度：39.8%
绿地率：30.2%

上海临港重装备产业区信息服务中心西邻黄沙港，建筑共五层，是以通信和信息机房、通信服务、商务办公为主的多功能综合性办公大楼。

设计理念为：由若干形体相同的长方形模块拼成一个方形，任意从中抽取一个模块，在保证其整体不会倒掉的前提下把抽出的模块放在整体的上方，以此类推继续下去。这是对于通信时代的思考得到的概念。模式化的东西可能存在于通信行业的方方面面，通过模式化的叠加，更加有效地简化人们的工作，为人们提供愈加轻松的环境。同时也要警惕他们所带来的潜在的隐患，其中任意一个环节的问题都会对全局产生严重的后果。

项目作为地块内开发的第一栋建筑，既要个性鲜明，同时要与地块南侧的已有建筑和谐。

建筑采用内院围合式的布局方式，建筑内部功能围绕内院布置，使得建筑具有良好的采光通风，也有效地提升了整栋建筑的生气。空中庭院将建筑内院上空放大，使内院不至于很狭隘，从空间感受上给人以更加宽敞舒适的感觉。

在首层设计了三个门厅——消费商务门厅、办公门厅和货物门厅，将不同的人流和货物流线分开。平面功能划分区域，各区相对独立，各层之间通过垂直交通相联系。

设计运用凹凸和虚实对比的手法来塑造形体。建筑以实体为主，虚体量主要体现在建筑的出入口处。由于建筑所处的位置关系，对建筑的屋顶（即建筑第五立面）也进行了精心的设计，屋顶的设计理念与整个建筑形式合而为一。

Shanghai Academy of Automobile Engineering Stage-II

Layout Design: French Di Birant
Architecture Design: Dblant Desgin International
Land Area: 58 000 m²
Floor Area: 85 960 m²
Floor Area Ratio: 1.48
Greening Ratio: 54%

Shanghai Academy of Automobile Engineering is a park centered on R&D, the design of which attempts to employ the simple and reasonable geometrical form as the basic architectural language to arrange the single buildings with varied requirements for the functions as well as their outdoor environment. In this manner, simple and smooth architectural space can be formed and highly effective working environment for R&D personnel can be built.

Buildings for testing: Large quantities of different testing workshops are based upon the requirements for the arranging process equipment. They are designed to be the reasonable and simple squares with relatively large dimension and high storey, which provides the façade windows with the degrees of freedom. The concept of science and speed are brought into the façade design. Both the size and the location of windows vary according to the inner lighting requirement, which creates a variable and dynamic effect. Face bricks of delicate texture are adopted for the walls to be in harmony with the buildings for R&D. The roof is designed as the 5th façade, and the wall elements are extended to the roof. Silver-grey rather than such bright colors as blue and red is chosen as the color for the roof board to display the unique gracious and scientific quality.

Buildings for R&D: The R&D building, modeling room and canteen are all people-oriented; therefore, the changing of unevenness as well as form of advancing and regressing are adopted on the basis of the shape of square. The sense of pressure brought by the massive buildings is loosen and small-sized squares and courtyards outdoors are established to satisfy the gathering and scattering of the stream of people and to enrich the architectural pattern with graceful disorder. When it comes to selecting the architectural styles of building for R&D, good academic atmosphere for automobile R&D mechanism is given full consideration. Bricks and stones are used as the material for the exterior wall to present the thickness and heaviness of the building, and glasses with different light perceptions are used to form the transparent surface. Dark grey and white are taken as the theme color to show the deepness and calmness of the scientific research institutions.

Places along Zhuxing River and Park Road have natural river course scenery and display the significant image of the park area; as a result, the massiveness of buildings and the contour control are stressed: from west to east, modeling room, the 1# and the 2# R&D building, canteen, the 3# and the 4# R&D building appear sense of rhythm through the evenness and the form of advancing and regressing of blocks, and then the whole building ultimately ends with the high-rise main building of the 4# R&D building, which builds up a complete picture of building group.

The buildings within the park area have an average height in general, which helps to show sense of wholeness. The 4# R&D building in the central part is elevated to become the peak of the whole park area. If seen from the direction of such main roads as Shanghai-Nanjing Highway and Cao'an Road, a symbolic building image is available for the entire park area.

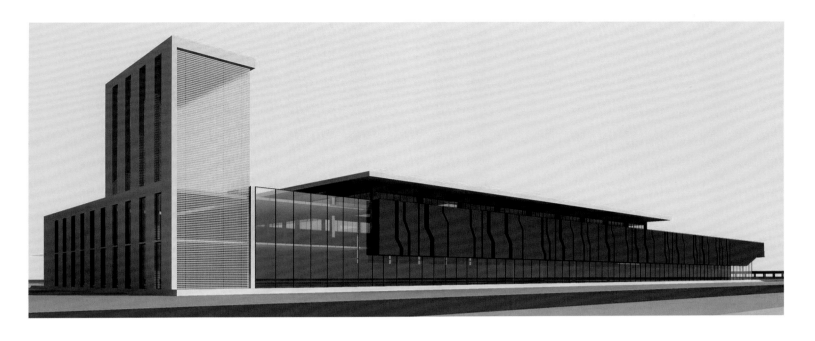

上海汽车工程院二期

规划设计：法国迪比兰特
建筑设计：都林国际设计
用地面积：58 000 m²
建筑面积：85 960 m²
容积率：1.48
绿地率：54%

上海汽车工程院是以研发为核心的园区，设计试图以较为单纯的、理性的几何形体作为基本的建筑语言，来组织具有不同使用功能要求的建筑物单体及其室外环境，从而形成简洁、流畅的建筑空间，为研发人员创造高效的工作环境。

试验类建筑：大量的各种试验车间，以布置工艺设备的要求为主，设计成理性的、简单的方形，尺度较大，层高较高，这为立面开窗形式提供了自由度。设计将科技、速度的概念引入立面设计，窗的大小和位置均因内部采光需求而不同，富于变化、具有动感。墙面用质感细腻的面砖，与研发建筑相协调。将屋顶作为第五立面来设计，把墙面元素延伸到屋顶。屋面板颜色选用银灰色，而不用蓝、红等鲜艳的颜色，以表现出高雅、科技的特质。

研发类建筑：研发楼、造型室、食堂是以人为主体，因此在方形的基础上，做出凸凹、进退的变化。削弱大体量建筑给人的压迫感，也在室外形成小的广场和庭院，满足人流的聚散，使建筑形态更丰富，高低错落有致。在研发楼建筑风格的选择上充分考虑汽车研发机构良好的学术氛围。外墙材料以砖、石体现建筑的厚重；以光感不同的玻璃来形成通透的表皮。以深灰与白色为主色调，体现科研机构的深沉冷静。

沿朱行河和沿园路既有天然河道景观，又是园区重要形象展示面。因此重点控制建筑体量、轮廓线：从西向东，造型室、1#2#研发楼、食堂、3#和4#研发楼通过体块的凹凸、进退，呈现出节奏感、最终以4#研发楼的高层主楼作为整组建筑的休止符，从而构成了一幅完整的建筑群形象。

园区建筑总体高度平均，表现出整体感。将中部4#研发楼拔高后，使之成为整个园区的制高点。从沪宁高速、曹安路等几条主要道路看过来，整个园区具有了标志性的建筑形象。

设计构思图

1,2#研发楼立面　　　3#研发楼立面

整体的建筑群体如何与沿河景观发生关系...

在建筑立面上开启"取景框"...

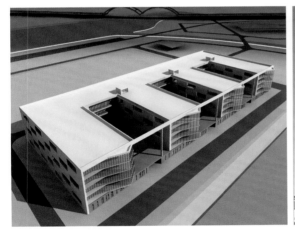
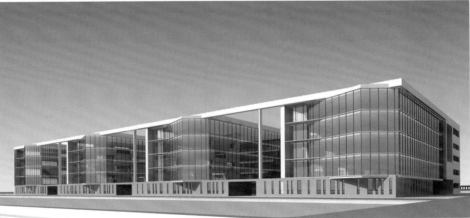

Guiyang North Zhonghua Road Reconstruction Project

Designed by: France AUBE Architecture and Urban Planning and Design Company
Land Area: 32.59 ha
Floor area: 2 320 000 m²
Floor area ratio: 7.12

Guiyang North Zhonghua Road Reconstruction Project is located in north of the central area of Guiyang City, adjacent to the administrative resource-intensive area of Guizhou Province and Guiyang, near bus terminal, railway station and other transportation facilities, as well as the Fountain, the Grand Cross and other well-known commercial circles in Guiyang, so the location is extremely advantageous.

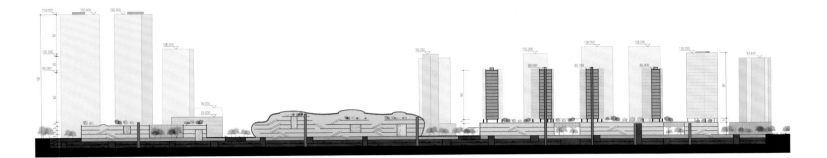

Overall layout

"One axis plus five cores" strip layout conforms to the city's texture relations, running through the principal axis in form of "green belt" as a landscape corridor themed with leisure, entertainment and health; the Central Forest Park as the focus of the project is located there; on the main line, the Conference and Exhibition Business District, Cultural District, National Fitness Zone, the Central Park Area, Religious and Cultural Tourism Zone are placed one by one from south to north to form the layout mode of "one axis plus five cores".

Landscape system

The natural and ecological forest park as the accessible and penetrative green core provides a quiet place for the urban platform in intensive exchange and forms a strong intentional character.

Transport system

The multi-level urban three-dimensional transport model: the urban vehicle flow line crosses under the platform, avoiding to the largest extent the disturbance to the internal lot; the pedestrian and emergency car-flow line is at the ground floor, creating a close-to-nature and environment-friendly living environment.

Commercial layout

Around-the-clock commercial area connecting the street and the corridor: in order to achieve balanced development of social and economic benefits, the success of commercial layout is the key element.

City node

Architectural image showing the spirit of times of New Guiyang: the entire style is combined with the city's character based on a concise and modern tone. With the use of simple and bright modern architectural design practices, it tries to show the internal function and spatial variation of the architecture.

贵阳市中华北路城市复兴项目

设计：法国欧博建筑与城市规划设计公司
用地面积：32.59 ha
建筑面积：2 320 000 m²
容积率：7.12

贵阳中华北路城市复兴项目位于贵阳市城市中心区北部，紧邻贵州省和贵阳市行政资源密集区，附近有客运站和火车站等交通设施，以及喷水池、大十字等贵阳著名商圈，地理位置十分优越。

总体布局

"一轴五核"的带状布局。顺应城市的肌理关系，以"绿带"的形式贯穿整个主轴，作为以休闲、娱乐和健康为主题的景观走廊，中央森林公园作为项目的重点位居其中；在主轴线上，由南到北依次布置会展商务区、文娱艺术区、全民健身区、中央公园区、宗教文化观光区，形成"一轴五核"的布局模式。

景观系统

自然生态的森林公园。通过通达、渗透的绿色核心，为密集交流的城市平台提供了安静的场所，并形成强烈的意向特征。

交通系统

多首层的城市立体交通模式。城市车行流线从平台之下穿越，最大避免了对片区内部的干扰。人行及紧急车行流线位于地面层，营造出贴近自然，环境友好的人居环境。

商业布局

街廊连接的全天候商业地带。为了做到社会效益和经济效益平衡发展，商业布局的成功是关键的一环。

城市节点

体现新贵阳时代精神的建筑形象。整体风格结合城市性格统一考虑，以简洁、现代为基调。运用简洁明快的现代建筑设计手法，力图表现建筑的内部功能及空间变化。

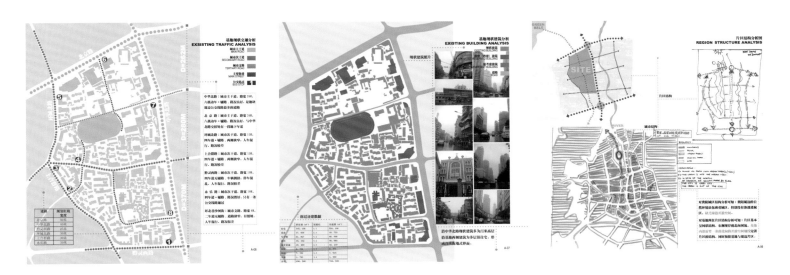
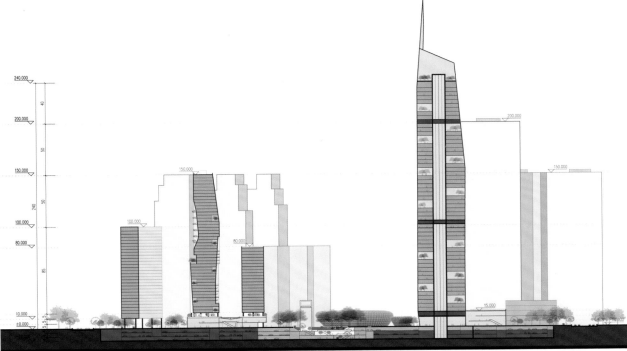

沿街立面
FRONTAGE FACADE

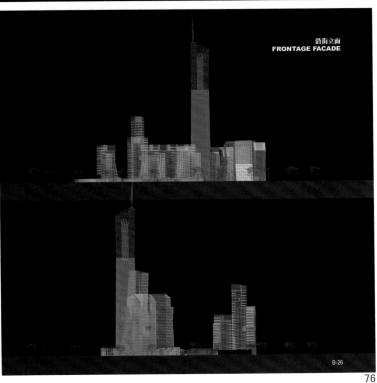

沿街立面
FRONTAGE FACADE

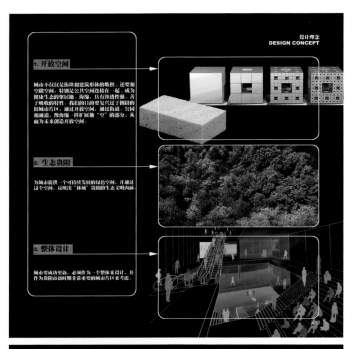
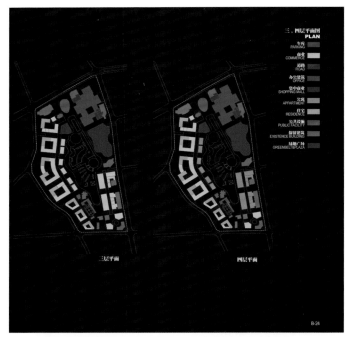
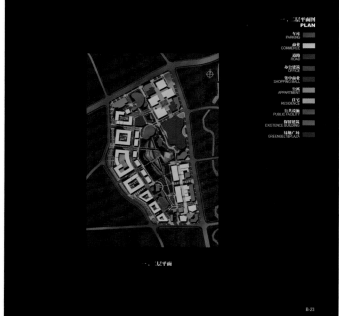
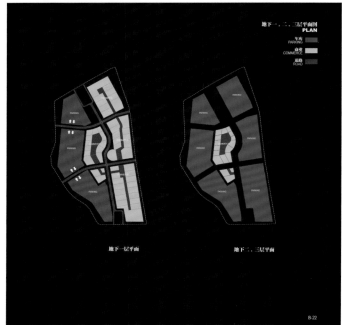

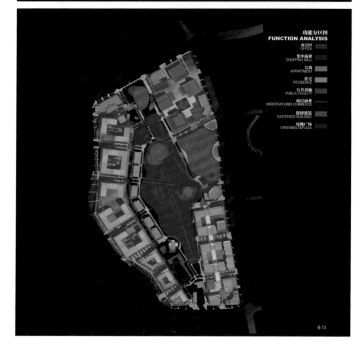

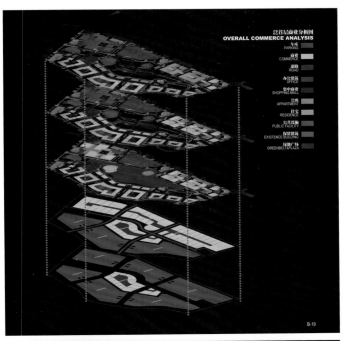
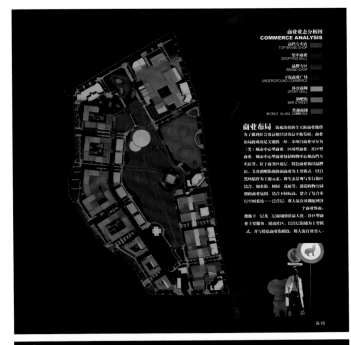
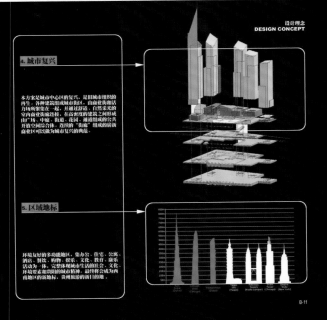
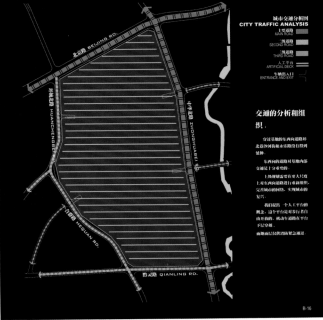
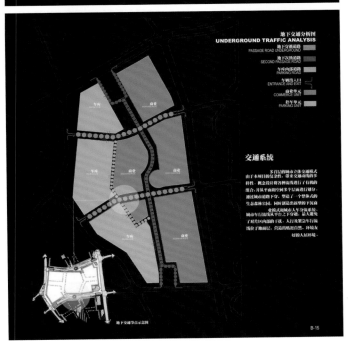
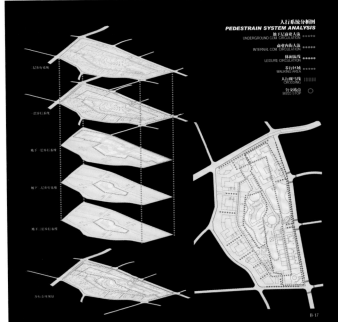

HOTEL ARCHITECTURE

酒店建筑

Front Coast Arts Hotel

Designed by: Shenzhen Zhang Miao Architectural Design Office

Front Coast Arts Hotel, the landmark structure of Baoan F518 fashion Creative Park, is a project not only based on the cultural industries, but also a project combined with old houses reconstruction meaning and performance of avant-garde art. "Alternative, avant-garde, pioneer, controversial ... also, the cost of investment has been limited." It has pioneering concepts and avant-garde visual impact ... abandoning the traditional design and regarding the "building skin" as an extension of interior space, rather than a separate so-called "facade design." Diversity of architectural interior space shall decide the overall shape to achieve harmony within and outside the space ...

There is none can be called "standard room" in the hotel! Because every room is full of change and personality: there are guest rooms for temporary residence, long-term lease rooms for SOHO, flat floor rooms, duplex leaping layers rooms, MINI studio at 30 to 40 square meters, workstation for over 200 square meters ... a total of 200 sets. This idea will also go on in future interior design ...just because of which, it greatly attracts countless designer friends who like self-expression of individuality.

1F ~ 6F is designed to be an integrated space for the hotel, which combines: Chinese and Western restaurants, banquet halls and bars and other catering services; also various types of meeting rooms and open-exhibition Hall designed for office, meetings, products promotion; multi-purpose halls prepared for press releases and theatrical performances; recreational fitness facilities and a small sunny swimming pool on 6F ... the most special site is the "peak" forum set on top floor of the entire building for designers, which is also a club for designers and a communicating place for variety of

cultural and artistic genres ... such a warm atmosphere is just the same as the building's shape – a pool of clear water filled from the top is slowly flowing down along the stone walls, wherever it goes could make a strong contrast to the bright smooth streams (architectural glass position) and stone walls ...streams of creativity source is flowing down, passing guest rooms, through the exhibition, into the F518 Creative Park ...

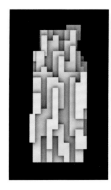 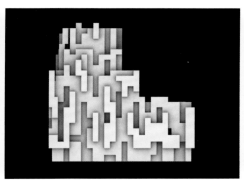 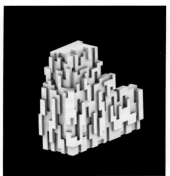 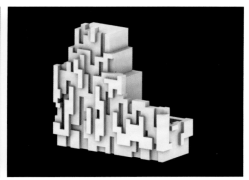

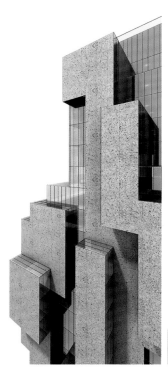  

前岸艺术酒店

建筑设计：深圳市张淼建筑设计事务所

宝安F518时尚创意园的标志性建筑前岸艺术酒店是一个立足于文化产业的项目，同时也是一个结合了旧房改造内涵和先锋艺术表现的项目。"另类、前卫、先锋、争议……还有，投资成本已经限定。"具有先锋概念和前卫视觉冲击力……摒弃了传统的外观设计，将"建筑外皮"视为内部空间的延伸，而不单独进行所谓的"外立面设计"。让多样性的内部空间决定建筑的整体造型，从而达到内外空间的和谐统一……

酒店没有一间可以叫做"标准客房"的房间！因为这里的每一间客房都是极富变化及个性的：有临时居住的客房，也有长期租用的SOHO；有平层房间，也有复式跃层房间；有30~40m²的MINI工作室，也有200多m²的工作站……总共200余套。而未来的室内设计也必将延续这个理念进行下去……也正因如此，它极大地吸引了无数喜欢表现自我个性的设计师朋友们。

1F~6F设计为酒店的综合用房，其中融合了中西餐厅、宴会厅及酒吧等餐饮服务内容，还有为办公、会议、产品交流推广而设计的各类会议室和开放式展览厅；也有为新闻发布及舞台表演而准备的多功能会堂；在6F还设置了娱乐健身设施及小型阳光泳池……而最特别的是整栋建筑的顶楼为设计师们安排的"高峰"论坛，这里也是设计师的俱乐部，是各种文化艺术流派"华山论剑"之所……这样热烈的氛围正犹如这栋建筑的造型一样——从顶部慢慢涌起的一池清水，顺着石壁缓缓流下，所到之处就形成了和石壁对比强烈的明亮光滑的溪流（建筑上玻璃的位置）……这一股股创作的源泉顺流而下，流过客房，流过会展，流入F518创意园……

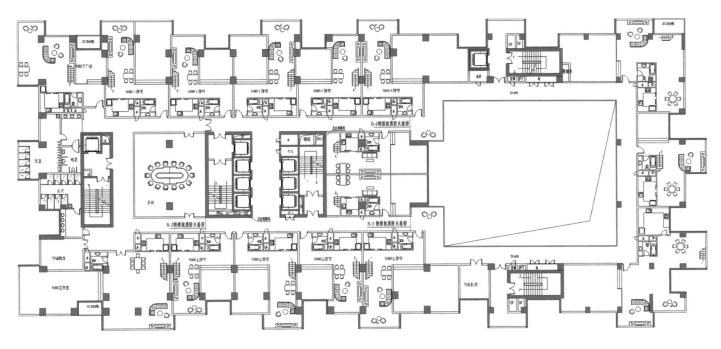

极具个性及多样性的SOHO工作室

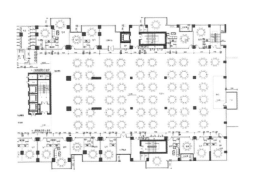

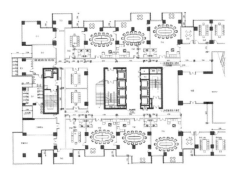

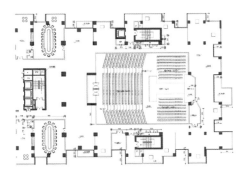

宴会厅　　　　　　　　　　　多功能会议　　　　　　　　　　大型会展

Wuxi HILTON Yilin Hotel

Designed by: Shanghai ZhongWaiJian Engineering Design and Consulting Co., Ltd
Cooperative Design: Japanese KKS Office
Main Designer: Zhaoyu, Wang Zuxu, Shen Jiaqi, Na Xiaomeng
Land Area: 25 178 m^2
Floor Area: 92 172 m^2
Floor Area Ratio: 3.0
Altitude: 100 m
Storey: 27

Wuxi HILTON Yi Lin Hotel is located in Wuxi New Area on the south of Gaolang Road, at the west of the fourth Hangchuang Road and by the east of Jianghai road. Main functions contain a five-star hotel, boutique shopping centers, class A office and luxury hotel-style apartments.

Church on the right and river on the left in the base has provided the environmental planning with a good cultural atmosphere and visual experience. The 25-storey main hotel building disposed along Gaolang Road and the 100 meters office building are joint together via a 4-storey commercial podium. These reasonable planned import

and export marks, direction signs of hotels, offices, apartments and commercial halls will ensure the orderly flow of traffic, passenger s and transportation division. Combination of the ground parking lot and underground machinery parking garage makes adequate parking space for future use.

The structure elevation adopts a blend design based on Jiangnan and modern architectural culture. Its symbol feature is enhanced by twin towers mass. The building seems elegant and forceful with the form of a clear body mass and a large number of vertical lines. The building's outer wall texture and horizon contour line are enriched by taking the advantage of material's organic changes and top part's collection or distribution. Contrast of the podium outer walls'

realty and fantasy, warmness and coldness shall a good help for the structure to impress itself, forming a good business climate. Entrance of the hotel lobby is metaphorically disposed like a city gate, which absorbs the verve of traditional and regional architecture to welcome guests and friends from around the world in the appearance of open and zest.

There is a large virescence courtyard behind the lobby outside the leading way of glass chamber, which is full of Chinese traditional gardening scenery charm. The high-rise folded setback style virescence is obstructed with the round miscellaneous sights to form a quiet garden for leisure. Perfect fifth façade is composed by the sky garden and tennis court on the Grand Ballroom's roof.

无锡HILTON逸林酒店

建筑设计：上海中外建工程设计与顾问有限公司
合作设计：日本KKS事务所
主要设计师：赵彧、王祖旭、沈佳奇、那晓萌
规划用地：25 178 m^2
建筑面积：92 172 m^2
容 积 率：3.0
建筑高度：100 m
建筑层数：27

无锡HILTON逸林酒店项目位于无锡新区高浪路南侧、行创四路西侧、江海路东侧。主体功能包括一座五星级酒店、精品商场、甲级写字楼及豪华酒店式公寓。

建筑基地右侧的教堂与左面的河流为环境规划提供了良好的文化氛围和视觉感受。沿高浪路布置25层的酒店主楼与100m高写字塔楼，两栋高楼通过四层商业裙房联为一体。合理安排酒店、办公、公寓、商业等出入口位置及导向标志，保证车流、人流、物流等各类交通分合有序。地面停车场与地下机械停车库上下结合，为建筑未来的使用提供充裕的停车位置。

建筑立面采用江南建筑文化与现代建筑风格的设计融合，通过双塔的体量增强建筑的标志性。通过形式清晰的体块构成，采用大量的竖向线条以增强建筑的峻秀挺拔。利用材质的有机变化及顶部的收与分来丰富建筑的外墙肌理与天际轮廓线。通过裙房外墙面的虚实对比和冷暖对比增强建筑的个性表达，形成良好的商业氛围。酒店大堂入口的处理则采用城市大门的喻意，吸纳地域建筑的传统神韵，以通达开放的面目欢迎来自五湖四海的宾朋。

酒店门厅的后方通高玻璃室外设置一个大型的绿化庭院，富有中国传统造园手法的对景意韵。采用退台叠高式绿化与南侧芜杂的现状视线阻隔，形成内向安静的休闲花园。大宴会厅屋顶布置空中花园和网球场，形成完美的第五立面。

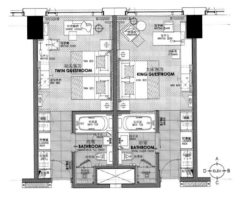
客房平面图

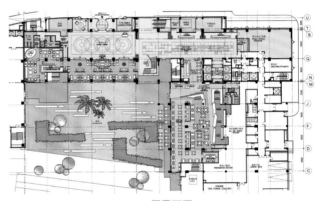
一层平面图

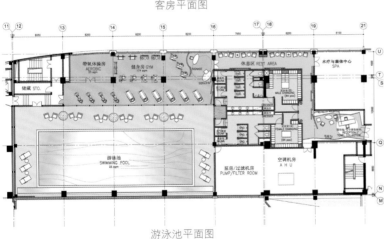
游泳池平面图

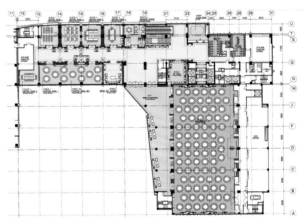
宴会厅平面图

Beijing Hongfu Hotel

Designed by Bohua Construction Design Office (HK)
Floor Area: 25 759 m²
Floor Area Ratio: ≤33.01

Beijing Hongfu Hotel project is located in the Sihuan Middle Road in the west of Haiding district. The project is convenient in traffic adjoining the west of Sihuan Middle Road to the west and Jingouhe Road to the south. Northern section of construction body is mainly for banquet, restaurant and office; southern section is living area. On this basic, the roof of the section for banquet is designed into a garden, and restaurant, banquet and office are closely combined to greatly improve the environment of the hotel. Between the construction body and Sihuan Road is set for greening. The space inside the yard is kept clear to melt into the surrounding environment of Sihuan Road. The design is rational, solemn but not dull, sprightly and stable and can deliver certain sense of times and characteristics.

Every part of the design is very important with abundant functions and vivid body; on the other hand, the banquet, restaurant and office section in the north is designed to set off the grandness of the reception hotel in the south.

The architecture consists of sections of different heights and they are as a whole by serving as a foil to each other. By adopting the open and transparent ternary form architecture processing method including the indentation at roof and the fenestration in the middle to facilitate the façade, perfectly embody the functions and be different from surrounding buildings in the image. The façade is processed by the combination of stone painted with nano-coating and glass. White stone and light blue glass are combined into one, together with gray grooved steel layering to make the building look stable as well as vivid. The partial use of black marble enables the building to melt into surrounding environment.

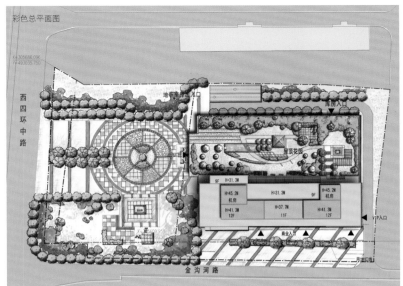
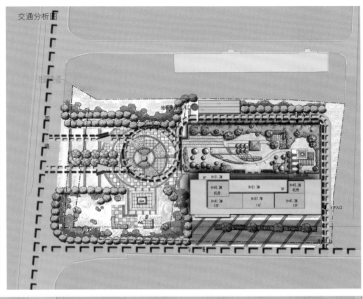
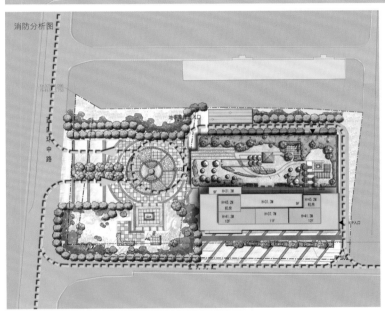
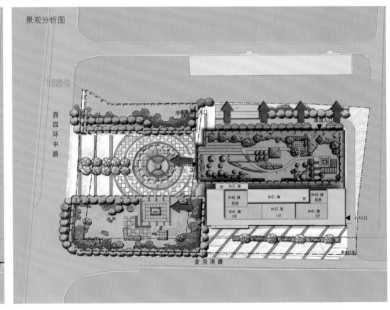

北京鸿府宾馆

建筑设计：博华建筑设计（香港）事务所
建筑面积：25 759㎡
容积率：≤33.01

北京鸿府宾馆项目位于北京市海淀区西四环中路，本项目西侧毗邻西四环中路，南侧为金沟河路，交通便捷。设计建筑单体的北部设计为主要宴会、餐厅、办公区；南部为宾馆住宿区。在此基础上，将宴会区屋顶设计为屋顶花园，把宾馆与宴会办公紧密结合，使宾馆的总体环境大大提升，在建筑单体和四环路路之间作为集中绿化。保持院落空间的清晰，与四环路的周边环境融为一体。造型设计富于理性，庄严而不呆板，明快而不失稳重，具有一定的时代感和特色。建筑设计的每一部分都十分重要，功能较为丰富，体型相对活泼；另一方面，作为北侧的宴会餐厅办公区，它努力衬托出南侧的招待所宾馆的雄伟和高大。

建筑体型高低错落、互相衬托又合为一体。通过采用包括顶层退台、中部开窗以及开敞、具有通透感的建筑三段式的建筑处理手法，便于建筑的外立面完整体现本身使用功能，亦在周围的现状建筑群中脱颖而出。建筑外立面采取石材及纳米涂料的墙体和玻璃虚实结合的手法，总体玉白色的石材和浅蓝色的玻璃相互映衬结合，辅以灰色的槽形钢压条，使建筑沉稳而又不失活泼建筑.外表面局部黑色大理石的使用，使得建筑本身融入周边环境。

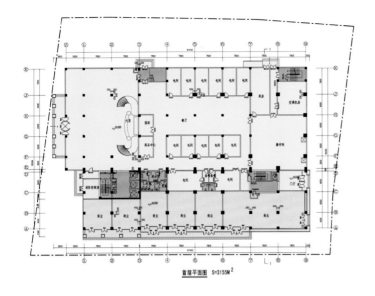

首层平面图 S=3155M²

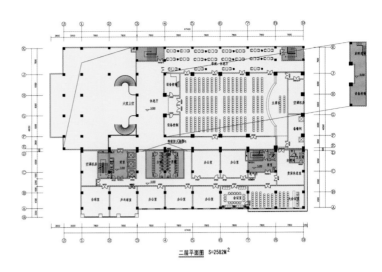

二层平面图 S=2582M²

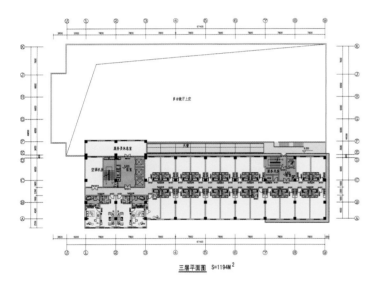

三层平面图 S=1194M²

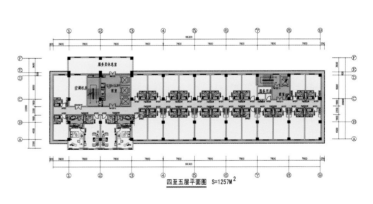

四至五层平面图 S=1257M²

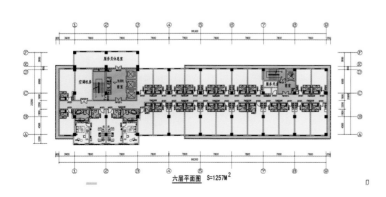

六层平面图 S=1257M²

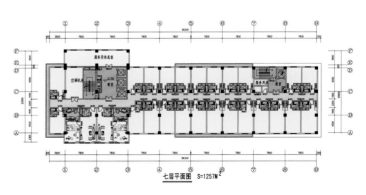

七层平面图 S=1257M²

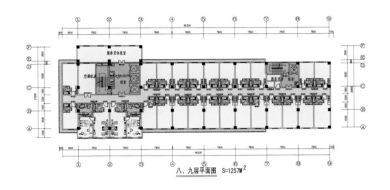
八、九层平面图 S=1257M²

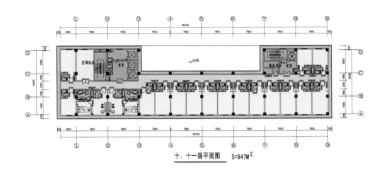
十、十一层平面图 S=947M²

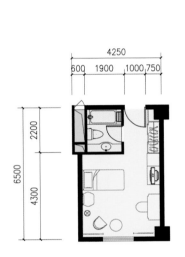
小标准间平面图

大标准间平面图

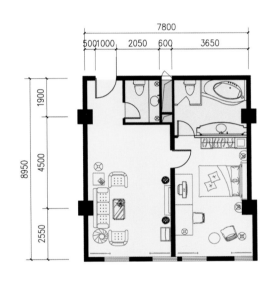
套间平面图

小总统套房平面图

大总统套房平面图

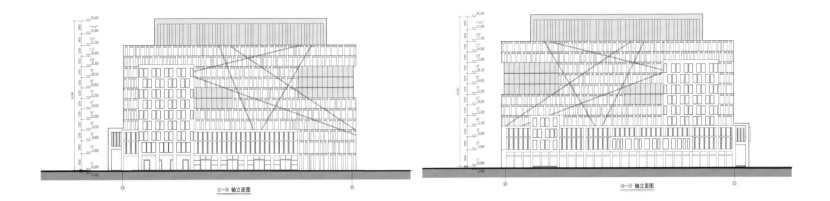

日景透视图（下）

Ganzhou Gannan (Holiday) Hotel

Designed by: Huasen Architecture and Engineering Design Consultants Company Limited
Collaboratively designed by: Hawaii DGH Designing Consultant Company

Ganzhou Gannan (Holiday) Hotel is a five-star hotel integrating business, tourism and hospitality together. It is developed and built by Ganzhou City Development and Investment Group Co., Ltd. and Jiangxi Bo Xin Real Estate Development Co.

The hotel covers an area of 94 361 square meters, applying the garden-style layout. There are 5 layers on the ground and 1 layer under the ground (half underground) with the building area of 55,559 square meters and 352 rooms equipped. Comfortable plane layout could maximize using the beautiful environment resources and combining with the natural landscape organically. It has mutual penetration, smart spiritual essence and profound artistic conception.

The 1433 square meters of multi-purpose hall, 380 square meters lecture hall and six medium-sized meeting rooms, as well as Chinese and western catering, bars, SPA, outdoor swimming pools, tennis courts and other facilities are all set up according to the five-star standard.

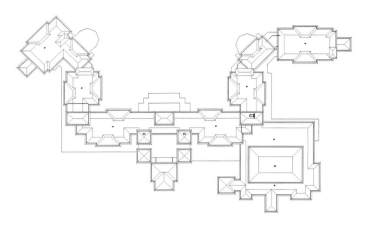

六层平面图

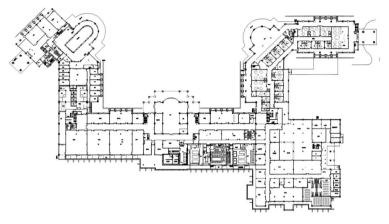

五层平面图

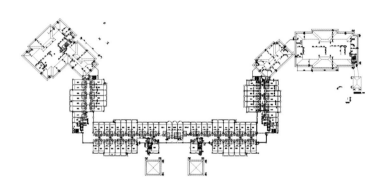

四层平面图

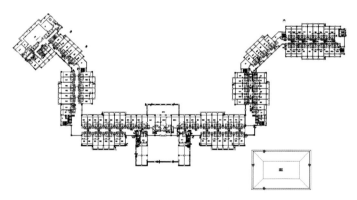

三层平面图

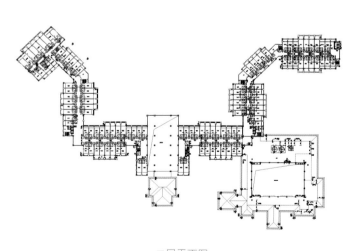

二层平面图

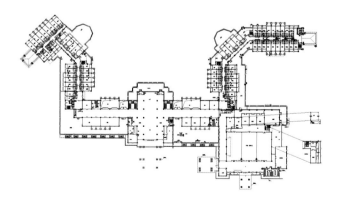

一层平面图

酒店立剖面

赣州赣南（假日）酒店

建筑设计：华森建筑与工程设计顾问有限公司
合作设计：夏威夷DGH设计顾问公司

赣州赣南（假日）酒店是一座集商务、旅游、接待为一体的五星级酒店，由赣州城市开发投资集团有限责任公司、江西博鑫房地产开发有限公司开发兴建。

酒店占地94 361m²，采用园林式布局，地上五层、地下（半地下）一层，建筑面积55 559m²，内设352套客房。舒适的平面布置最大限度地利用秀丽的环境资源，与自然景观有机组合，互相渗透，气蕴灵动，意境深邃。

1433m²的多功能厅，380m²讲演厅和6个中型会议室，以及中西餐、酒吧、SPA、室外泳池和网球场等设施齐全，都是按五星级要求配套设置。

Foshan Zongheng Hotel

Location: Foshan
Net Land Area: 18 506 m²
Floor Area: 68 244 m²
Floor Area Ratio: 2.5
Altitude of Architecture: 120 m
Parking Space: 410
Number of Guest Room: 500

Designed by: Huasen Architecture and Engineering Design Consultants Company Limited
Designers team: Qimeng Hu, Fan Wu, Hui Ren, Xiao-Yan Xie, Liang Yuan, Jiancheng Zhang, NINA

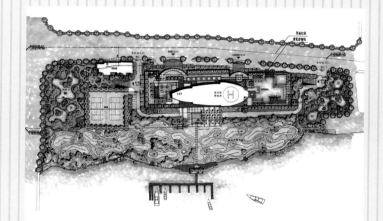

The future development goal for Foshan is to build the city into a modern metropolis with nice environment, which is suitable for living, investment and tourism. In order to coordinate with the International Ceramics Expo, the owners decided to invest and build a first-class standard five-star hotel in Foshan with domestic distinctive features, integrating business reception and meeting together. During the Expo, the hotel is mainly responsible for the business and conference, some

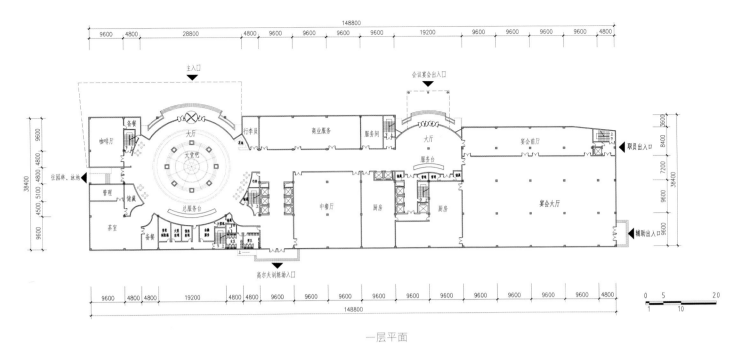

一层平面

functions will be changed for office use when it finishes. The site of the project is located by the side of Jili river levees in Foshan, where the waterscape could be enjoyed on the south, just watching the Kwan-yin effigy on Xiqiao Peak in the South China Sea 8 km away from the southwest outside. There is absolute landscape advantage while standing among the surrounding urban low-rises.

The hotel lobby and conference hall are designed in parallel, forming the double-hall structure, in which there is a super-luxurious two-storey circular large space in hotel lobby. Two lobbies are connected by a spacious commercial street inside. The distribution design and vertical traffic for flow of people from each part ensure the functional line to be convenient and simple. Lobby, conference, catering, entertainment and other various functional areas may provide comprehensive in accordance with the international five-star standard.

There is a unique spindle-shaped plane and full-arc façade for the tower to enlarge the viewing angle of each guest room and bring the pleasant waterscape from out of the windows as far as possible. Two separate luxury 270° landscape suites are standing on each end of the standard layer, providing guests with a full range of urban vision. In order to adapt the function transformation after Expo, the design adopts standard column network structure and builds the luxury 9 floors of suites to be 4.5 meters high for each storey which sit above the refuge layer.

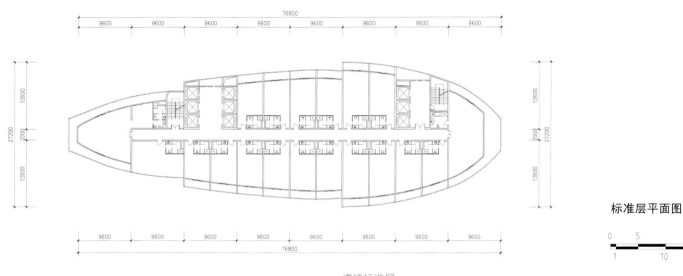

塔楼标准层

标准层平面图

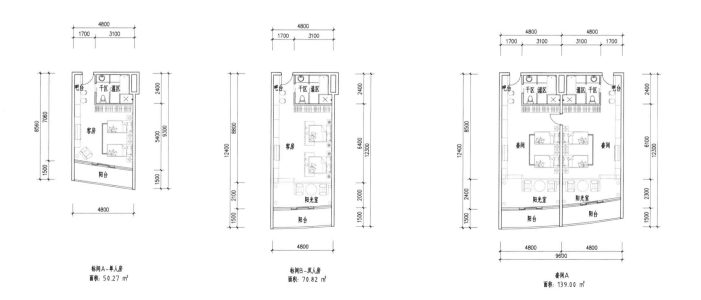

标准间A-单人房
面积: 50.27 m²

标准间B-双人房
面积: 70.82 m²

套间A
面积: 139.00 m²

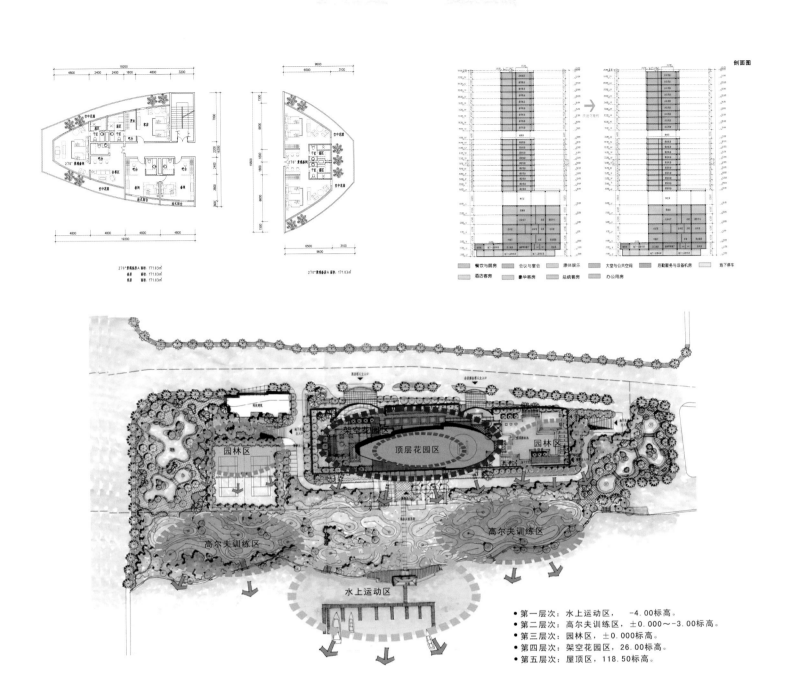

佛山纵横大酒店投标

项目地址：佛山
净用地面积：18 506m²
总建筑面积：68 244m²
容积率：2.5
建筑高度：120m
停车位：410
客房数：500

建筑设计：华森建筑与工程设计顾问有限公司
设计团队：胡起萌、吴凡、任辉、谢晓燕、袁亮、张建成、NINA

佛山市未来发展目标是建设一个环境优美，适宜居住、投资、旅游的现代化大都市。为了配合国际瓷器博览会，业主决定投资建设一座佛山市第一、国内有特色的，集商务接待和会议一体的标准五星级酒店。在博览会期间，酒店主要负责博览会的商务和会议需求，博览会结束后，有可能会将部分功能改做办公之用。项目用地位于佛山吉利河涌防洪堤边，南面坐拥水景资源，正好遥看西南8km外的西樵山顶南海观音像，在周边低矮的城市建筑群中，具有绝对的景观优势。

本案并列设置了酒店大堂和会议大堂，形成双大堂型结构，其中酒店大堂为超豪华的两层高圆形大空间，两个大堂通过一条宽敞的商业内街联系。各部分人流结合竖向交通，进行分流设计，保证了功能流线的便捷和单纯。大堂、会议、餐饮、娱乐等各个功能区都能按照国际五星级标准，提供完备的服务。

客房塔楼采用独特的梭形平面，弧度饱满的外立面，能够放大每间客房的视野角度，尽可能大地收揽窗外的宜人水景。标准层两端，各有一套豪华的270°景观套房，为客人提供全方位的城市视野。为了适应博览会以后的功能改造，设计采用标准柱网结构，并将避难层以上的9层客房，设计为4.5m层高的豪华套房层。

Sanya "Beautiful Crown"

Designed by: Huasen Architecture and Engineering Design Consultants Company Limited

The original Beautiful Sanya Crown Culture Exhibition Center was completed in December 2003. As the first meeting hall for the 53rd, 54th and 55th Miss World finals, the oval-shaped main building is in steel membrane structure (106M × 92M). i The "Beautiful Crown" is particularly enchanting under the night curtain, which has already become a shining pearl in people's hearts in Hainan.

Four members of Huasen Hotel Designing and Researching Group came to this picturesque place for a field trip commissioned by the new owners in July 2007, who were seeking for another brilliant idea to make the "Beautiful Crown" more beaytiful.

Base of Sanya "Beautiful Crown" covers an area of 237,000 square meters, to the west of Phoenix Road and at the east of Sanya (eastern) River, bank of which has lush mangroves comparing with the beauty of famous flowers park in the north. According to design requirements, there is a super five-star hotel, demonstration hall, business clubs, business matching facilities and other projects to be built. The conceptual design has just begun and an enforceable design program of "glory recreation" is in the pipeline.

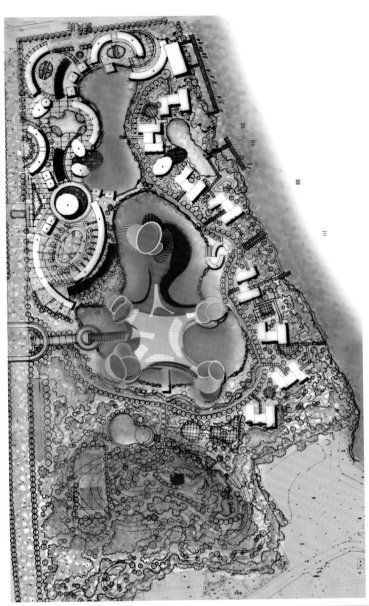

三亚"美丽之冠"

建筑设计：华森建筑与工程设计顾问有限公司

原三亚美丽之冠文化会展中心2003年12月建成，作为第53、54、55届世界小姐总决赛会场，椭圆形的主体建筑采用钢构膜结构（106m×92m），尤其是夜幕下的"美丽之冠"分外妖娆，已经成为海南人民心中的一颗璀璨明珠。

2007年7月接受新业主的委托，华森酒店设计研究组一行四人来到这块风景如画的地方实地考察，为"美丽之冠"更加艳丽、再创辉煌寻求新的创意方案。

三亚"美丽之冠"基地面积237000m²，东临凤凰路，西临三亚（东）河，河岸红树林郁郁葱葱，北面名花公园更是秀丽。按设计要求这里要建设超五星级酒店、演示会场、企业会所和商业配套等项目，概念设计刚刚开始，一个可实施的"再创辉煌"的设计方案正在酝酿中。

Zigong Zhangjiatuo Totel

Designed by: Huasen Architecture and Engineering Design Consultants Company Limited

Zhangjiatuo Hotel is the first five-star hotel in Zigong. There are 250 guest rooms with the construction area of 50 000 square meters. It also has the five-star standard matching facilities for catering, conference, leisure and gym. The public part of the hotel is spreading along rive coving according to its geographical features of being surrounded by rivers and mountains, facing river bays. The approaching water platforms and pedestrian system are set up, which create a quiet, beautiful urban environment together. The hotel appears a gesture of reclining the mountain due to its mountainous landform on the back side, which gives guests the feeling of living between rivers and mountains, enjoying the nice scenery. At the same time, the hotel will bring a new bright spot to the city, lead the reconfirmation of the whole region and provide new developing opportunity.

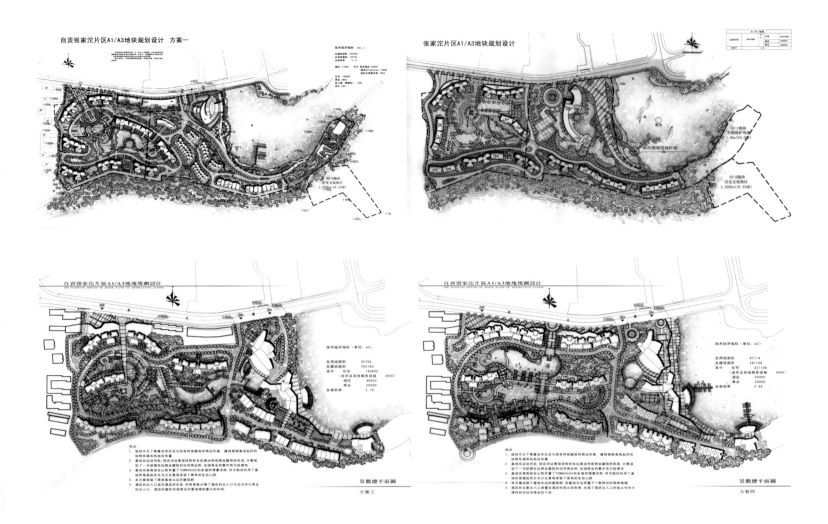

自贡张家沱酒店

建筑设计：华森建筑与工程设计顾问有限公司

张家沱酒店是自贡市第一个五星级酒店，设有客房250间，建筑面积50 000m²，配套有五星级标准的餐饮、会议、休闲及健身设施。酒店规划设计结合依山傍水的地形，面临河湾的地理特征，将酒店公共部分倚河湾展开，设亲水平台和步行系统，在都市中创造一个宁静优美的环境。酒店又利用背后的山地地形，形成倚山之势，使酒店住客深深感受到身居山水之间，保证了客房拥有良好的景观。同时，酒店的建设给城市面貌带来新的亮点，并带动环境的改善，带动整个区域的重新整合，带来新的发展机遇。

Conceptual layout for phase three of Huangshan Songbai Golf Hotel

Designed by: Huasen Architecture and Engineering Design Consultants Company Limited
Area: 57 659m^2
Scheme Designer: Liu Xin

Conceptual layout for phase three of Huangshan Songbai Golf Hotel is a five-star hotel which contains townhouse-style resort hotel, holiday villas and 202 suites. There are three forms in the design: a five-star hotel, townhouse-style hotel apartments, divisibly- used villas. The hotel's phase three expansion and townhouse-style hotel apartments have strong continuity with the plots for phase one and two, highlighting the integration layout for the overall group in the design. Based on the comprehensive analysis of the site patterns and functions, it shall start with the business hotel rooms (Building A) and apartment-style rooms (Building B) of phase three in the design, it would be the business hotel rooms part which is close to the phase one hotels, while the apartment-style rooms part is in the most peripheral site and extending into the plot of townhouse hotel-style apartments. Townhouse-style hotel apartments are mainly in modular splicing way and an appropriate enclosure in the more free plate form. In addition, land for villa construction is relatively complete, so it is designed as a more independent region. There is a vast villas area sandwiched by the fairways on both sides. It applies the island layout in the design, each "island" is relatively independent to be a small-scale garden resort hotel; "islands" are connected by garden roads and landscape corridors. While it looks like a huge house inside the "island", which could accommodate a large number of independent living units. Each unit is in the apartment-style interior layout. Site of phase three Songbai Golf Hotel is hills with natural gentle slope, the building program has fully respected the existing topography, arranging each groups of structures in the hotel community in accordance with local conditions, which not only has smooth connections, but also maximizes to protect the existing topography.

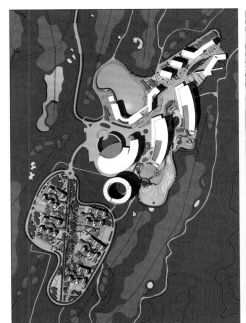

黄山松柏高尔夫酒店三期概念性规划

建筑设计：华森建筑与工程设计顾问有限公司
面积：57 659m²
方案设计人：刘新

　　黄山松柏高尔夫酒店三期概念性规划含联排别墅式度假酒店、度假别墅等，202间套房，为五星级酒店，设计分三种形态：五星级酒店、联排形式的酒店式公寓、可分割使用的别墅。酒店的三期扩建、联排式酒店公寓，与一、二期用地连续性强，在设计中突出体现群体的整合规划。根据对用地形态以及功能的综合分析，在设计中选择将三期酒店的商务客房(A栋)、公寓式客房(B栋)进行，靠近一期酒店的是三期的商务客房部分；在最外围的是公寓式客房部分，并且延伸进联排酒店式公寓的用地范围内。联排酒店式公寓主要采用单元拼接的方式，以较自由的板式形态作适当的围合。此外，用于别墅建设的用地相对较完整，设计中作为较独立的区域规划设计。被夹在两侧球道之间一个较开阔区域的别墅区，在设计中采用岛式布局，每个"岛"相对独立，成为一个个小型的花园度假酒店；"岛"与"岛"之间以园林路以及景观廊加以连接。"岛"的内部像一个大宅院一样可以容纳若干独立的居住单元，单元内拥有公寓式的室内布局。松柏高尔夫酒店三期用地形态为自然的缓坡丘陵，建设方案充分尊重现有地形，因地制宜的安排酒店群落内的各组建筑，做到既有顺畅的连接，又最大限度地保护原有地形、地貌。

SHOPPING MALL

商业广场

Kunshan Boyue Piazza

Architecture Design: Shanghai Domestic and Foreign Construction Engineering Design and Consulting Co., Ltd
Collaborative Design: Japanese KKS Office
Leading Designer: Zhao Yu, Wang Zuxu and Fuwen
Planned Land Area: 41,005 m²
Floor Area: 34,122m²
Floor Area Ratio: 3.27
Building Altitude: 100m
Stories: 27

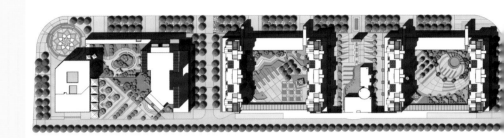

Kunshan Boyue Piazza project is distributed in the southwest of Kunshan Economic and Technical Development Zone, east of Bailu Road and north of Zhonghua Road. The base is like a long and narrow rectangle with the distance from south to north measured 400 meters and east to west 100 meters, together with elegant surrounding environment, convenient traffic and open view. Many landmark and memorable projects are under construction around, such as Kunshan Catholic Church, Bailu Park and Jitian Piazza which will occupy 700 thousand square meters.

The development consists of two stages. The stage-I project in the north can be also divided into two clusters. For each cluster, between two 18-storey apartments is set with business matching facilities; the roof is a shared garden for residents of apartments; two-lane car ramp extends to the roof and connects to the entrance of each unit without people's awareness of any traffic barrier or loss of sight caused by non-ground factors. Residents may walk to the hall of public chamber between the two clusters to take an escalator directly up to roof garden and then enter the hall of unit. This design truly separates resident flow from business and guarantees the convenience and privacy of residents. With the technique of 3-D space, designers set comparatively independent flows for business people, vehicles and cargos to create a business atmosphere. Meanwhile, complete business facilities also ensure the convenience and life quality of residents.

The stage-II project is positioned at the reinforcement and upgrade of the stage-I. The large-sized business and catering services provided by the 4 floors of podium building will add to the popularity and prosperousness of the area. The two towers are for offices and hotel-style apartments. Its large mass, handsome and upright appearance and thick modern atmosphere are mixed to raise a visual climax of the whole architecture cluster.

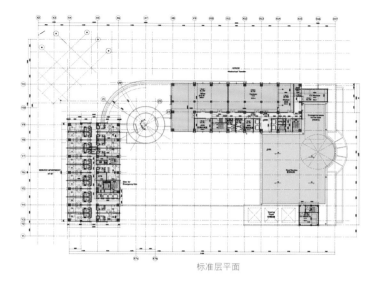
标准层平面

主立面

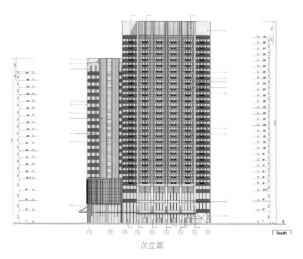
次立面

昆山博悦广场

建筑设计：上海中外建工程设计与顾问有限公司
合作设计：日本KKS事务所
主要设计师：赵戎、王祖旭、付文
规划用地：41 005m²
建筑面积：134 122m²
容 积 率：3.27
建筑高度：100m
建筑层数：27

　　昆山博悦广场项目位于昆山经济技术开发区西南，柏庐路东侧，中华路以北。基地南北约400m，东西约100m，呈矩形狭长地块，周边环境优美，交通便利，视野开阔。昆山天主教堂、柏庐公园、70万m²的吉田广场等一大批具有标志性和纪念性的项目在周围开工建设。

　　开发进程分为南北两期，北面的一期又分为两个组团，每个组团两栋18层公寓之间设置商业配套，商业屋面是公寓居民共享的花园绿地，双车道的汽车坡道可直通屋面并与各单元入口连接，感觉不到非地面导致的交通障碍和景观缺失。步行的居民还可以从两个组团之间的公共会所大堂乘自动扶梯直接上到屋顶花园进入单元门厅，真正意义上做到了商业与居住人流的分离，并保证了住户生活起居的便捷与私密，以立体空间的组织手法形成商业人、车、货等流线的相对独立完整，使商业氛围得以实现。同时，完备的商业设施又保证了住户居民的生活便利与质量。

　　二期的开发定位是一期的功能补充与提升。四层裙房的大型商业与餐饮服务将带动区域的人气与繁华，两座塔楼分别是办公与酒店式公寓。体量高大，俊朗挺拔，浓郁的现代气息掀起了整个建筑群的视觉高潮。

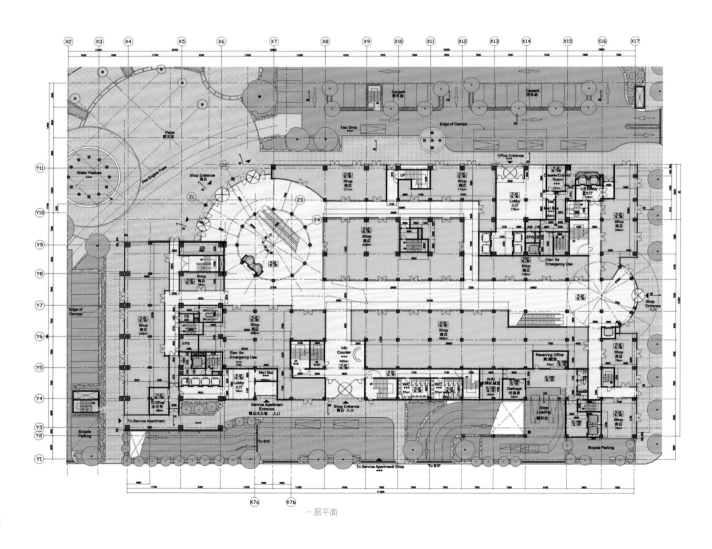

一层平面

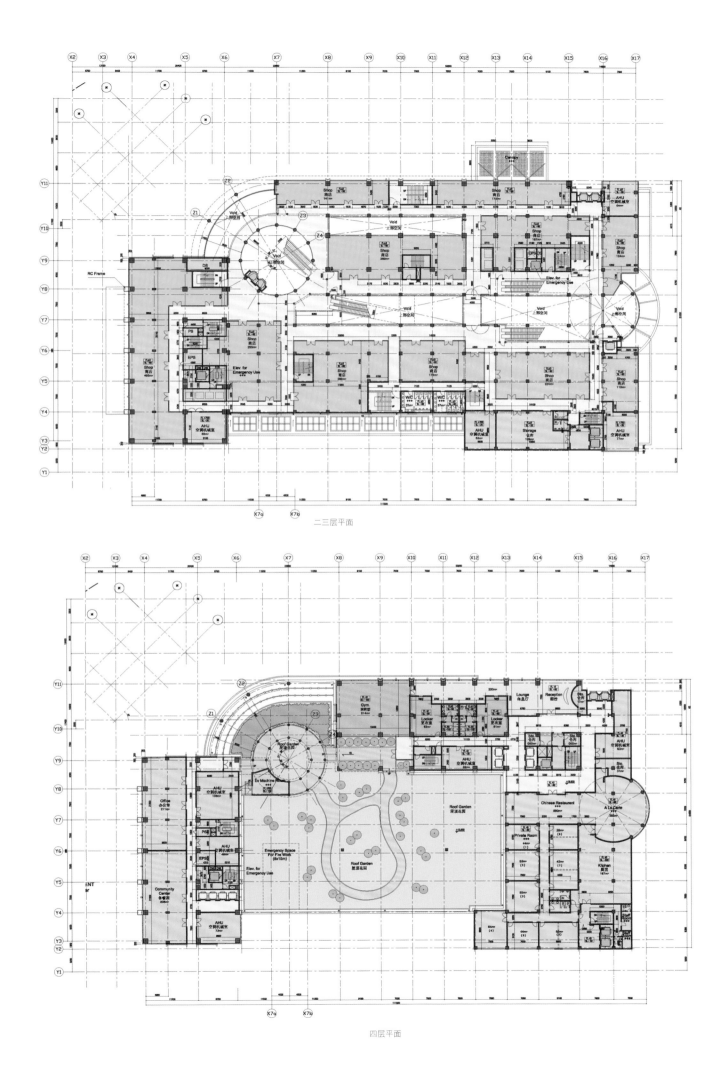

二三层平面

四层平面

Wuyang Commercial Plaza

Designed by: Shanghai K.O.E Architecture Design Office

Land Area: 23 3225 m^2
Floor Area: 91 305.6 m^2
Floor Area Ratio: 3.06
Building Density: 34%
Greening Ratio: 12%
Parking Space: 414

Base of Wuyang Commercial Plaza is on the south of Shanghai Road, at the north of pending office area, by the east of riverway and to the west of Dongcang Road, including three parts of apartments, high-rise office buildings and economical hotels. There is full consideration to the morphological characteristics of important buildings around for the overall layout design. The spatial relations of urban nodes shall be perfected from perspective of urban design. The office building is placed on the northwest side of plot, standing towards the Century Square office building on the north of Shanghai Road. Business is settled along Shanghai Road and Dongcang Road, main entrance of sidewalk is set in the meddle part of the road, which may lead the flows of people from the northern street and eastern street into the base, connect the eastern main entrance with the business portal interface in the east of Dongcang Road as well. High-rise office building is located in the southern plaza with southern entrance settled, while there is a way connecting with the north entrance of bank. Main entrance of the principal commercial stores is set up towards the northern end of the street, while the entrance of economical hotels is set up towards Dongcang Road, exits and entrances for cargo flow and logistics are set up towards the loop part of apartments on the north. One pedestrian street and an inner courtyard are taken as the transportation center of commercial portion, where people can walk through the western arcade and extend business line until the commercial zone in front of the river. Two buildings of apartments are designed to found on the south end

of plot, respectively for 24 and 22 layers with independent traffic flow lines, entrances and exits. Apartments and commercial houses have independent entrances and exits, south of which are standing against commercial zone to avoid the interference.

Office building's plane and vertical composition is striving to legitimately lead the flow of people. The suitable space proportion, good lighting and airiness, comfortable office environment create a modern architectural space charm together as the commanding point of the land. Commercial complex is designed on the concept of linear business + plate business. Facade of the main office building takes the classical ART-DECO style to create a noble elegance of the neo-classical European charm. Details stress the shadow relationship of lines and planes combination, appropriate placement of the advertisement could avoid a chaotic pattern of confusion.

Apartments are in full lighting design, square-framed and practical with clear functional flow lines, specific dynamic and static partition without any wasting space. Design of residential facade emphasizes the coordination of whole offices and commercial pattern, trying to be concise and dignified, avoiding interference with the business offices. The outside facade takes the classical style as blueprint, highlighting the refinement and delicacy.

五洋商业广场方案设计

建筑设计：上海高亚建筑设计事务所
总用地面积：23322.5m²
总建筑面积：91305.6m²
容积率：3.06
建筑密度：34%
绿地率：12%
停车位：414

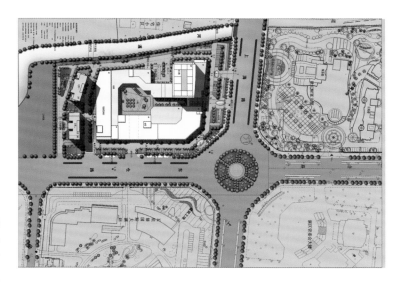

五洋商业广场基地北为上海路，南邻待批办公区，西临河道，东为东仓路。含公寓、高层办公楼、经济型酒店三部分。设计总体布局充分考虑周边重要建筑形态特征，从城市设计的角度入手，完善城市节点的空间关系，办公楼置于地块西北侧，与上海路北侧世纪广场办公楼相呼应。沿上海路及东仓路布置商业，在路中部设人行主入口，即将北街商业人流及东街商业人流引入基地；又考虑使东主入口与东仓路东街商业入口衔接。高层办公楼设南广场并设南进口，北设道路连通银行北进口。商业主力店面向东北端马路设主入口，面向东仓路设经济型酒店主入口，商业北面面向公寓环路处设货流及后勤出入口。商业部分以一条步行街及一个内庭院作为交通中心，并可信步穿过西过街楼，使商业动线延伸至面河商业一带，形成点、线、面的丰富空间模式；围绕庭院作内廊，自动扶梯及电梯连通各个楼层，形成富有活力的共享空间。庭院内可举办各种商业活动，同时与步行街相通，形成了顺畅的交通流线。商业的南侧布置经济型酒店，较为安静，出入口与商业分开。地块南端设计两栋公寓，分别为24及22层，有独立的交通流线及出入口。公寓与商业出入口独立设计，南侧背向商业，主要使用房间不受商业干扰。

办公建筑的平面和竖向构成，力求人流疏导合理，空间比例适宜，采光通风良好，办公环境舒适，作为该地块的制高点，力求营造现代建筑空间魅力。商业建筑群按线性商业+板块商业为理念进行设计。办公楼主体立面采取古典式ART-DECO风格，营造高贵典雅之新古典欧式风情，外立面采用高档仿石面砖，局部点缀石材，既简洁大

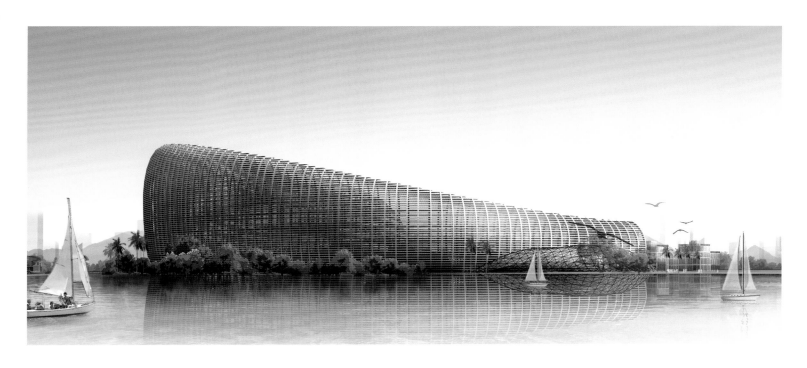

方又不失精致。商业部分以高档石材穿插玻璃幕墙为主，衬托大型商业设施的庄重、气派，不会随时代的变迁而落后，并自然形成了入口的宽敞感。细部处理强调点线面组合的阴影关系，并妥帖安置广告位，避免形成杂乱无章的混乱格局。

公寓全明设计，户型方整实用，功能流线清晰，动静分区明确，无浪费空间。充分利用每一平方米面积，将小户型房型达到精巧细致、功能齐全，又舒适宜人的整体效果。住宅立面设计注重与整体办公、商业格局的相互协调，尽量简洁、端庄，并避免对商业办公的干扰，外立面亦以古典风格为蓝本，彰显精致、细腻。

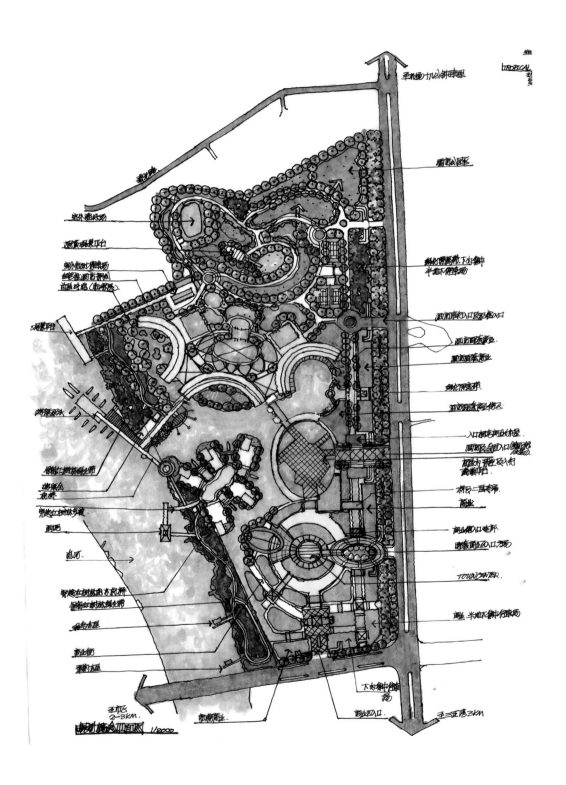

Commercial Architecture on Dongting Highway Block No.4 of Songjiang District of Shanghai

Designer: Japan RIA Research Institute of Architecture

Carry on the old Shanghai's culture and spirit
During the process of carrying on the architectural style, the design pays attention to the spirit conveying, pursues "accuracy", "aesthetic quality" and "fun", through innovative design creates distinctive style, great taste and shape unique language with the leisure space highlighting the beauty of details all over to reflect the petty bourgeoisie's elegant life of old Shanghai.

New business core with complex functions
Business function develops towards diversity, which by introduction of art galleries and other cultural facilities and integration of commercial and cultural functions enhances the overall style and creates a harmonious commercial culture to meet the demand for leisure and shopping of people at different age levels, suit both refined and popular tastes and keep pace with the contemporary commercial tastes so that Shanghai Customs Street in the future will become a paradise for shopping and leisure and create a business culture guiding the trend.

Fun shopping experience
Now the consumer behavior has changed that people have no longer a clear purpose for shopping but take shopping as a way of entertainment, a kind of leisure-based and experience-based consumption and supplement to the consumption form in residential area or in the future. People have been integrated into the business atmosphere to fully experience the shared and dramatic life which becomes the place for friends and family gathering and ideal homeland to enjoy a comfortable life.

Inheritance and Innovation
The design summed up the architectural style of Old Shanghai – British style, American style, southern China-style, mixed style and so on, trying to show the classics of Old Shanghai while integrating new forms and new materials for both cultural heritage and innovation, making it the product of combining the old and new. Based on combination of Western and Chinese culture, the design provides consumers with a unique way of life and consumption form from the old street and advocates a leading leisure culture trend to achieve the "new elegant Shanghai-style life" in real sense.

The introduction of natural and ecological environment
The sunshine and green is set in the atrium and balcony, etc., and is introduced indoor to form ecological artificial climate. In view of not large depth in the commercial architecture the design tries as far

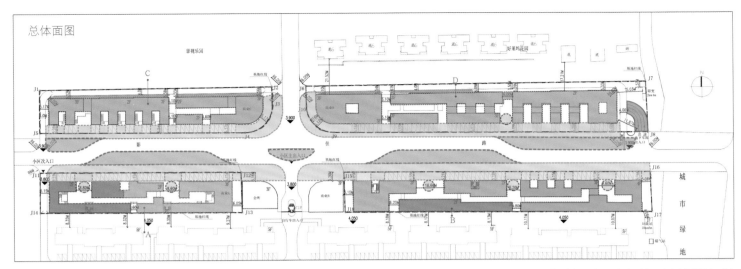

总体面图

整个商业部分由4各部分组成：A、B地块由商业、文化、娱乐、广场、中庭空间等组成，其中文化设施包括画廊等，娱乐休闲设施包括夜总会、咖啡厅、酒吧、美容美发、足浴、茶楼和棋牌室、服饰店等。商业设施包括零售商店、超市、宾馆等。C、D地块主要是主力餐饮和特色餐饮构成，包括中餐、各类小吃、西餐、日本料理等。

as possible to realize north-south ventilation and lighting to reduce energy consumption, such as air-conditioning; at the outdoor public space you can experience at any time seasonal and climate changes, enjoy the sunshine, breathe the fresh suburban air and listen to the current sound and pleasing music.

Organic combination of various commercial spaces
The commercial architectural plan integrates atrium, plaza and other leisure facilities inside, making the space real or virtual, dense or sparse and full of change to maximize the functional use of commercial space and win the greatest economic benefits. Focusing on the courtyard, combined with inner street, overhead building, balcony and other architectural elements of old Shanghai, the design forms enclosed or semi-enclosed space so that the external and internal space become more closely linked and people at shopping or leisure time can shuttle freely, which is full of fun.

立面图

立面图

立面图

立面图

上海松江区车亭公路四号地块商业建筑

设计：日本（株）RIA都市建筑设计研究所

老上海文化与精神的传承

在建筑设风格传承的过程中，注重精神的传达，打造商业建筑设计的"精度"、"美度"和"趣味度"，通过设计上的独具匠心，创造风格鲜明、品味不凡、塑造出自身的个性化语言，处处彰显细节之美的休闲空间，体现老上海小资的优雅生活。

复合功能的新型商业核心

商业功能向多元化发展，将画廊等文化设施引进，商业与文化功能的融合，提升了整体的格调，营造和谐的商业文化，满足不同年龄层次人们休闲购物需求，体现雅俗共赏，把握时代商业品味，使未来的上海风情街成为购物休闲的天堂，打造出引领时尚的商业文化。

体验购物乐趣

现在人们的消费行为已经发生变化，逛街不带有明确的购物目的，是作为一种休闲娱乐的方式，是一种休闲型消费、体验式消费，是对居住区或未来人们消费形式的补充。

人们融入商业氛围中，充分体验生活的共享性和戏剧性，成为亲朋好友聚会的所在和享受舒适生活的理想家园。

传承与创新

设计中总结了老上海的建筑风格－英式、美式、江南风格、混合式风格等，在力图表现老上海经典的同时又融入了新形式和新材料，力求在文化上得到传承和开拓创新，是新旧结合的产物。以中西融合为基调，为消费者提供一种老街独有的生活与消费方式，倡导一种领先于社会前沿的休闲文化潮流，实现真正意义上的"优雅海派新生活"。

将自然生态环境引入

阳光与绿色设置于中庭与露台等处，并将其引入室内，形成生态的人工气候，商业建筑的进深不大，尽量实现南北通风和采光，以减

少空调等能源的消耗；当置身室外，随时体验季节变迁和气候幻化的公共空间，享受阳光、呼吸郊区的新鲜空气，聆听水流声和悦耳的音乐。

各类商业空间的有机结合

商业建筑规划将中庭和广场等休闲性设施融入内部，使得空间有实有虚，有疏有密，富有变化，达到商业空间在功能上最大化的利用，并得到最大的经济效益。以庭院为主，结合内街、过街楼、露台等老上海建筑元素，形成有围合、半围合的丰富空间变化，使得外部和内部空间的联系更加密切，人们在购物休闲时可以自由穿梭，富有趣味性。

Suzhou Jing-Hang Canal Entertainment Center

Designed by: ANS International Architecture Design and Consulting Co., Ltd
Designer: TONY, ANDY

Land Area: 15 375 m²
Floor Area: 5 928 m²
Base Area: 15 375 m²
Greening Area: 7 730 m²
Storey: 2-3 stories
Floor Area Ratio: 0.39
Greening Ratio: 51%

This magnificent landmark building displays the amalgamation of grid structure and curve or beeline in vertical space.

As a fashion center, the building may combine social intercourse with the public facilities. Modern restaurants, cafes, nightclubs, KTV box, fashion shops, fine galleries and so on shall be all embodied in this internationalization building, which is also closely linking to context of Suzhou.

The design has taken full consideration of its exact location, Canal Road on the west, Beijing-Hangzhou Grand Canal by the east,

Dengwei Road to the north and Jinshanbang Riverway at the south. The traffic position importance is also taken seriously after detailed surrounding landscape check, landscape around which has formed a whole body with its geographical resources.

Interaction of lighting, water elements and architectural forms would be very critical to this architectural design with highly identity.

Design has wisely considered the contrast between reality and fantasy space. Traditional grid, as the chief structure, combined with vertical closing elements may highlight the characteristics of the building.

苏州京杭运河娱乐中心

设计：ANS国际建筑设计与顾问有限公司
设计师：TONY、ANDY

用地面积： 15 375 m²
建筑面积：5 928 m²
基地面积：15 375 m²
绿化面积：7 730 m²
建筑层数：2-3层
容积率：0.39
绿化率：51%

这座宏伟的地标建筑展示了普通的建筑结构网格与竖向空间中的曲线、直线的交融。

建筑作为一个时尚中心，将社会交往与公众设施结合起来。现代餐厅、咖啡馆、夜总会、KTV包厢、流行服饰小店、精致画廊等都在这样一座国际化建筑中得以体现，与苏州的文脉紧密相连。

建筑的设计充分考虑了项目所在用地，西侧为运河路，东侧为京杭大运河，北侧为邓蔚路，南为金山浜河道。设计对周围的交通的景观条件周详地考虑了区位的重要性。地块周边区域的景观以及地理资源形成一个整体。

灯光、水系要素和建筑构造形式的相互影响对于这座具有高度识别性的建筑设计而言非常关键。

设计智慧地考虑了虚实空间的对比。传统网络作为主要结构与垂直闭合要素结合起来，突出了该建筑的特性。

立面图

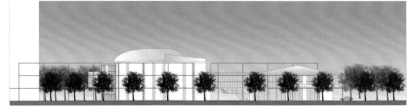
立面图

Shanghai Jiujiu Piazza

Planned and Designed by: ANS International Architecture Design and Consulting Co., Ltd
Designer: Liuli, Huangfei and Yuan De Yun

Land Area: 93 240m²
Floor Area: 205 778m²
Floor Area Ratio: 1.74

Shanghai Jiujiu Piazza project is at the Jiuting Station of Shensong Line of Shanghai Subway. The design will make the station a convenient, important and landmark power center and lively network center by connecting transfer central station, subway station, business, star-level hotel and office together through business corridor, in order to turn it into a multifunctional city complex driving the integral development of Jiuting district.

The design will try to create a well-arrange, elegant and modern construction cluster. Around subway station will construct the buildings with different functions to maximize the business value of constructions on the ground. To construct a landmark building in Jiuting area is the target of design. Meanwhile, the design will attach importance to the advantages along Puhuitang River and make it become the model of waterside constructions.

上海九久广场

规划/建筑设计：ANS国际建筑设计与顾问有限公司
设计师：刘莉、黄飞、袁德溁

用地面积：93 240m²
建筑面积：205 778m²
容积率：1.74

上海九久广场项目为上海的轨道申松线九亭站。规划将地铁站设计成便捷、重要、标志性的区域动力中心和活跃的网络中心。设计通过商业廊道将交通换乘中心车站，地铁站，商业，星级酒店、办公等空间串联，使其成为功能复合的都市综合体，带动九亭地区的整体发展。

设计力争创造一个布局完善、环境优雅、具有现代气息的新时代生态建筑群。规划概念围绕地铁站本体，建造不同功能的建筑，使其地块上层的建筑商业价值最大化，并以建造九亭地区城市地标性建筑为设计目标。同时注重发掘地块沿蒲汇塘河的区位优势，使之成为滨水建筑的典范。

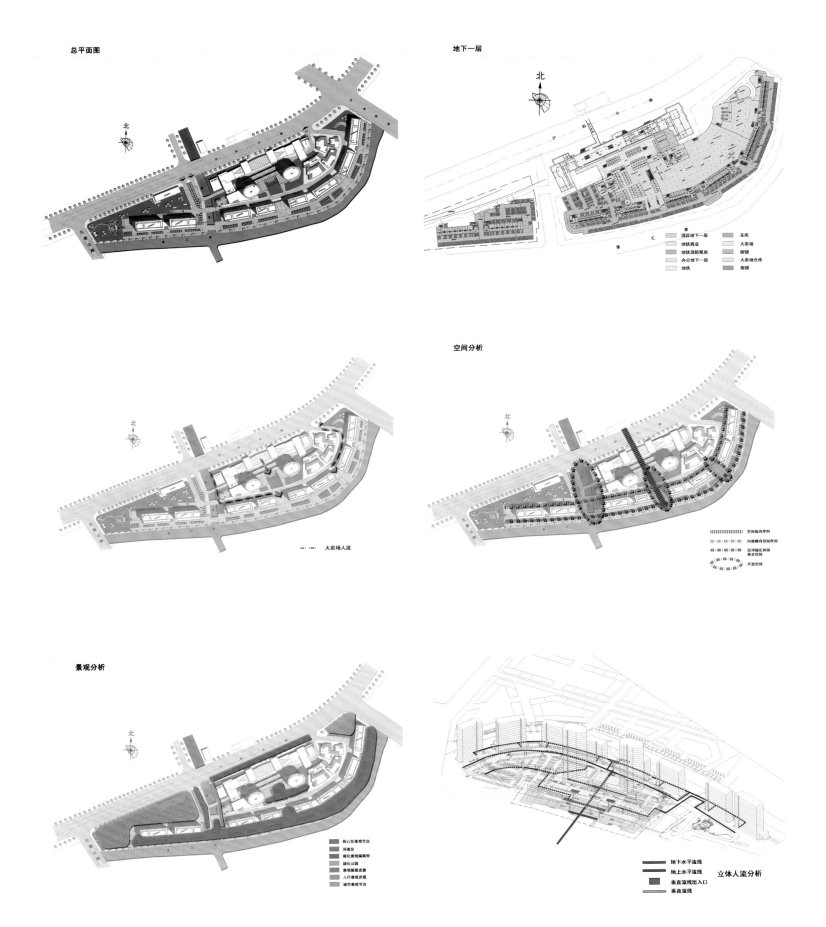

Wuxi Yiju International Life Piazza

Architecture Design: Shanghai Zhongwaijian Engineering Design and Consulting Co., Ltd
Major Designer: Xiaying, Wangliang, Luo Xianjun
Planned Land Area: 26 344.5 m²
Floor Area: 83 515 m²
Floor Area Ratio: 2.5
Architecture Altitude: 99.4 m

The principle of design is to follow the harmonization between architecture and environment and shape the elegant appearance through exquisite and reasonable design. While outstanding the unique temperament of architecture, design must meet the demands of urban planning trying to construct a modern space. The integral design must take the convenience of construction and demands for sustainable development into consideration, as well as the energy-saving and cost-effective demands under reasonable operation.

无锡逸居国际生活广场

建筑设计：上海中外建工程设计与顾问有限公司
主设计师：夏莹、王良、罗贤君
规划用地：26 344.5 m²
建筑面积：83 515 m²
容 积 率：2.5
建筑高度：99.4 m

设计原则为遵循建筑与环境的和谐，通过细致、合理的设计，达成优美的建筑外观。即突出本建筑特有的气质，同时符合城市规划的要求，努力营造一个富有现代气息的建筑空间。整体设计必须考虑到既要满足建筑物在使用功能方面的便捷性和可持续发展的要求，同时要满足在合理运营的前提下尽可能节约能耗、降低运行成本。

Nanjing Wanda Piazza

Location: Nanjing
Architecture: United Design Group
Land Area: 11.52 ha
Floor Area: 686 thousand square meters

Under the layout principle of wholeness and cultural continuity creates a modern and ecological urban open space. The design is to build a human-oriented urban space basic on two kinds of experiences, and an effective and comprehensive 3-D urban traffic system.

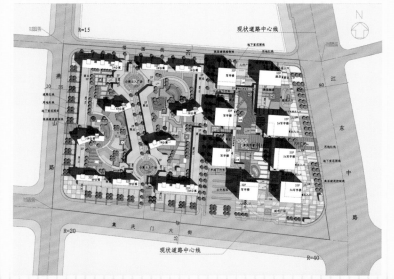

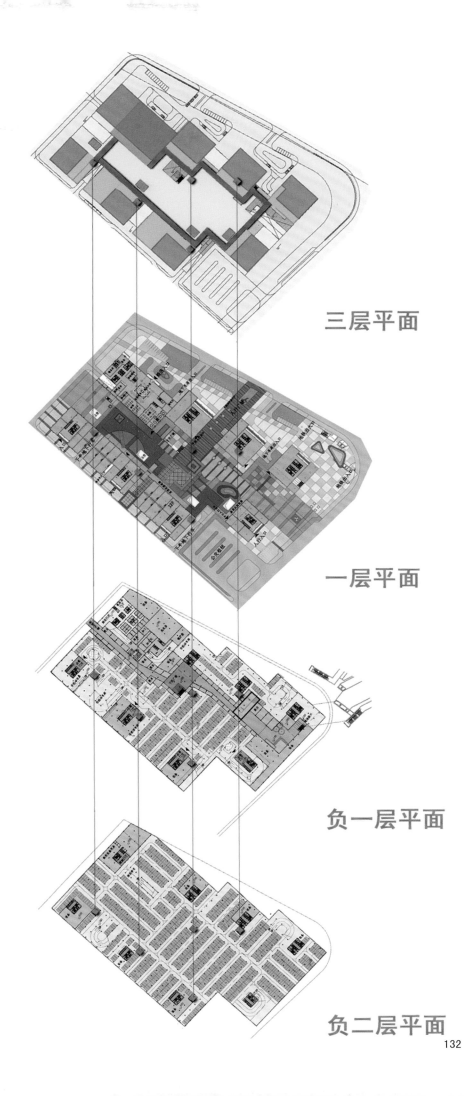

南京万达广场

建设地点：南京
建筑设计：UDG联创国际
用地面积：11.52ha
建筑面积：68.6万m²

在整体性，文脉延续性的规划原则下，创造现代化的生态城市开放空间。建立完善两种体验尺度下的人性化城市空间。高效综合的城市立体交通系统。

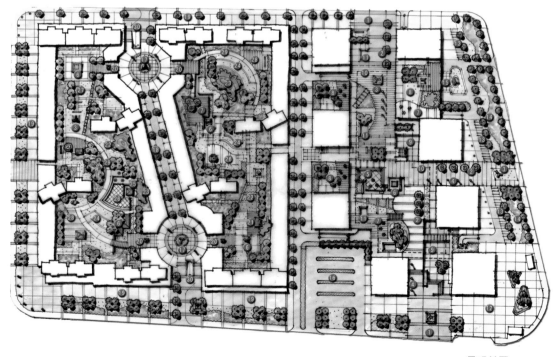

景观总图

Wuxi Wanda Plaza

Location: southwest of the intersection of Qingqi Road and Liangxi Road, Binhu District, Wuxi City
Designed by: United Design Group
Land Area: 179 300m²
Floor Area: 696 400m²
Floor Area Ratio: 3.0

Wuxi Wanda Plaza project is located in the central commercial area of Helukou in Lakeside region, which is on the west of Qingqi Road, to the south of Liangxi Road, at the north of Liangqing Road and by the east of Lixi Road. The plot is divided into five blocks, A、B、C、D、E, mainly for business, office and residence. Block A is planed for a high grade hotel, block B is used as a combined large scale commercial plaza of apartments, public leisure, entertainment and shopping, while blocks C、D、E are mainly for residence. Construction area of the project is 179,300 square meters and the overall construction area on the ground is about 537893 square meters.

Complex pattern of Wuxi Wanda Plaza contains an international large scale shopping center with an area of 230 000 square meters, one super platinum five-star hotel covering 40 000 square meters, one hotel-style apartment for 30 000 square meters, a high grade residence region with an area of 365 000 square meters, two intelligentized, ecological, human-natured, multiplex commercial buildings, one five-star Wanda International Cinema with multiple halls and one leisure plaza for citizens with the area of 30 000 square meters.

In order to adapt the local environment, climate and coordinate with local architectural culture, structure applies the light and transparent modeling. Meanwhile, the local modeling process of balconies and the floating windows is used to form its own unique design style, which forms Wanda's special construction format.

无锡万达广场

项目地址：无锡滨湖区青祁路和梁溪路交叉西南角
建筑设计：UDG联创国际
占地面积：179 300㎡
总建筑面积：696 400㎡
容积率：3.0

无锡万达广场项目位于滨湖区河埒口中心商务区，东至青祁路，北至梁溪路，南到梁青路，西为蠡溪路，用地分为A、B、C、D、E五块，主要功能为商业、办公、居住。其中A块规划为一处高档酒店，B块规划为精装公寓和集休闲娱乐购物为一体的大型商业广场，C、D、E块则以住宅为主，项目可建设用地面积约17.93万㎡，地上总建筑面积约537 893㎡。

无锡万达广场复合的形态包括：一个建筑面积23万㎡的国际化大型购物中心；一个建筑面积4万㎡的超白金五星级酒店和一个3万㎡的酒店式公寓；一个建筑面积36.5万㎡的高档住宅区；两栋智能化、生态化、人性化、多元化的商务楼；一个五星级的多厅万达国际影城；一个3万㎡的市民休闲广场。

建筑以轻盈通透的造型来适应当地的环境与气候，并与当地的建筑文化相协调。同时利用阳台、飘窗的局部造型处理，形成自身独特的设计风格，形成万达特有的建筑形式。

Tung Ying Building

Location: 1-19A Granville Road, Tsim Sha Tsui
Site area: 3 125.6 m²
G.F.A.: 37 507.2 m²
Developer: Chinese Estates Holdings Ltd

It's not easy to convince shoppers to seek retail therapy inside a high rise, but in the new Tung Ying Building in Tsim Sha Tsui the answer has been found in adhering to the principle, "Form follows Function". To attract maximum patronage, the building has been split into two blocks that will contain restaurants, shops and cinemas.

The circulation of the three-level podium is along the facades on both Granville Road and Nathan Road. From this, shoppers can either enter the 26 storey main tower or the 11 storey cinema block. The first 14 floors of the main tower are devoted to retail, and the higher floors are allocated to food and beverage outlets. The top four floors are set back creating terraces with great views of Kowloon and providing great spaces for al fresco dining. A long escalator connects the 6th floor of the main tower to the base of the cinema block, providing a second access point between the two blocks.

东英大厦

地点：香港尖沙咀连威老道一至十九A
用地面积：3125.6m²
总建筑面积：37 507.2m²
发展商：华人置业集团
建筑设计公司：梁黄顾设计顾问(深圳)有限公司

设计说明：要吸引善变的消费者到一幢高楼建筑中购物并不容易，但尖沙咀新东英大厦的设计师，以遵照"形式紧随功能"的原则，找到令人满意的答案。为了尽量吸引人流，特别将大厦分成两座，当中包含餐厅、商店及电影院。

综合项目位于加连威老道及弥敦道之间，人流可从三层高的平台前往26层高的主厦或11层高的戏院大楼。主厦的购物商场一直伸至14层，再上则为餐厅。由于露天用膳越来越受欢迎，因此特将大楼的最高四层移后，腾出空间建设阳台，让用膳者可眺望九龙以至更远的繁华景色。一条长长的扶手电梯由主厦的六楼一直通往电影院大楼底部，为两座大楼提供了另一个链接。除了视觉和实质上的连接外，电梯亦透视了整个综合体里的内部动态。

Wuxi New Urban Business Piazza on Plot C4

Architecture Design: Shanghai Zhongwaijian Engineering Design and Consulting Co., Ltd
Major Designer: Xiaying, Wangliang, Han Shichang
Planned Land Area: 60 099.5 m^2
Floor Area: 141 633 m^2
Floor Area Ratio: 1.79

The design is going to build the New Urban Business Piazza into an urban business park in order to achieve a commercial target in ecological way and set up landmark cognition with a special image. In the creation of social value, ecological value and business value, the design will also achieve a comparatively high level and make them supplement each other and develop in a harmonious environment. "Adjacent to landscape" is a vivid description we have for the image of the architecture. The body of architecture and the structure of green roof make the architecture an ecological peak, hillside, canyon and cliff. Various man-made ecological environments provides the city with a bright landmark images and the inside of these images also strictly abides by various principles of unique functional division of commercial building, flow line organization and introduction of flow, so as to improve the business value of each part and truly achieve various business targets.

无锡C4地块新城商业广场

建筑设计：上海中外建工程设计与顾问有限公司
主要设计：夏莹、王良、韩世昌
规划用地：60 099.5 m²
建筑面积：141 633 m²
容 积 率：1.79

　　设计方案将新城商业广场打造成一个城市中的商业公园，以生态的方式达成商业目标，以独特的形象建立地标认知，在社会价值、生态价值和商业价值的创造上，同时达到相当的高度，相辅相成，和谐共生。"山水相邻"是我们对于建筑形象的生动描述，建筑体型和绿色屋面的组织使得建筑在城市中的形象成为一座生态的山峰，山坡，峡谷，悬崖，各类人工生态环境为城市提供了鲜明的地标形象，而这些建筑形象又在内部严格符合商业建筑独特的功能分区、流线组织和人流引入的各种原则，提升各部分建筑体量的商业价值，真正实现了在商业上的目标："四方辐辏"。

MUSEUM, ART GALLERY AND EXHIBITION HALL

博物馆、艺术馆、展览馆

Design Scheme of Ai Weiwei's Works

Designed by German S.I.C-Engineering Consulting Limited Liability Company
Designer: Stephan Jasper
Works Style: Conceptual Design Creativity

"Module" is the works exhibited outside the exhibition hall in the German Kassel Documenta. Ai Weiwei built it with over 800 wooden windows of the styles of Chinese Ming-Qing dynasty like a specially shaped arch. This facility rising like a Gaudi Architecture brings people secret and dreaming symbolic experience. However, the works collapsed in the period of exhibition from the attack of a strong wind. Mr. Ai refused to rebuild it thinking that "it is destined to go through this integral process of coming from the ruin and back to the ruin." After viewing the ruin, Mr. Jasper got the idea of combining ruin with architecture and then started to make the design and draw the design sketch.

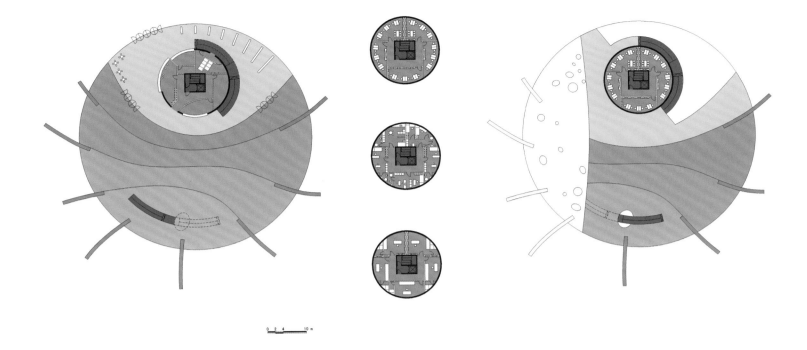

艾未未作品的设计方案

设计单位：德国S.I.C.-工程咨询责任有限公司
设计师：史蒂芬·亚斯伯（Stephan Jasper）
作品类型：概念设计创意

《模块》是德国卡塞尔文献展Kassel Documenta上放置在展览馆室外进行展出的作品。艾未未用了八百多块中国明清风格木门窗搭建而成，像一座造型特异的拱门。这一矗起如高迪建筑的装置，给人以神秘梦幻的象征性体验。但在展览期间，由于作品遭到大风的侵袭而坍塌。艾先生拒绝重新搭制，认为"它来自废墟，又化为废墟，天意地完成了一个完整的过程"。亚斯伯先生在参观过废墟后，产生了将废墟和建筑相结合的想法，于是动手做出了建筑方案设计并绘制了效果图。

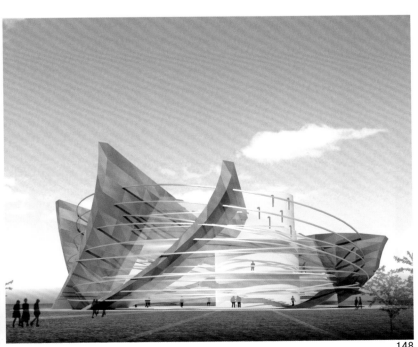

Franconia Jewish Museum

Franconia Jewish Museum
Location: Franconia
Designed by: German S.I.C.-Engineering Consulting Limited Liability Company
Designer: Jan Gutermuth and Kennan Domas
Land Area: 800m²
Floor Area: 1 260m²

The historic architecture "Franconia Jewish Museum" will be expanded through a new multifunctional construction. This project was set for an international competition and 30 design companies joined the competition. The result was announced by the jury on July 25th, 2008 and 3 companies won the prize and 2 companies got excellence award. The design of S.I.C. was granted the excellence award.

The new construction embodies the profound background and long history of Judaism in the Franconia. In figuring the urban sight of Fuerth (a city in German Bavaria state), designer adopted the material from local sandstone. The architecture is very similar to the sandstone monolith which is the representation of the culture of Franconia and is divided by the obvious piling way of sand layer. The new museum architecture and academy museum both play a similar and symbolic role.

Obviously, the "piling of sand layer" stands for the accumulation over the time. It lets people get to know and understand the long history of the Jewish and their influence in Franconia. Like a geologist or archaeologist, it lists, analyzes and expatiates on the main functions of museum one by one. From the angle of implied meaning, the historical segment of Franconia will be shaped into a sandstone body piled up by sand layers, which is what we call as surface rock, and the body will be raised and stand in the urban landscape.

Dialogue: Old-New
"The most important exhibition" in the Franconia Jewish Museum is its own historic architectures. This background also helps explain the style and direction of this new architecture body.

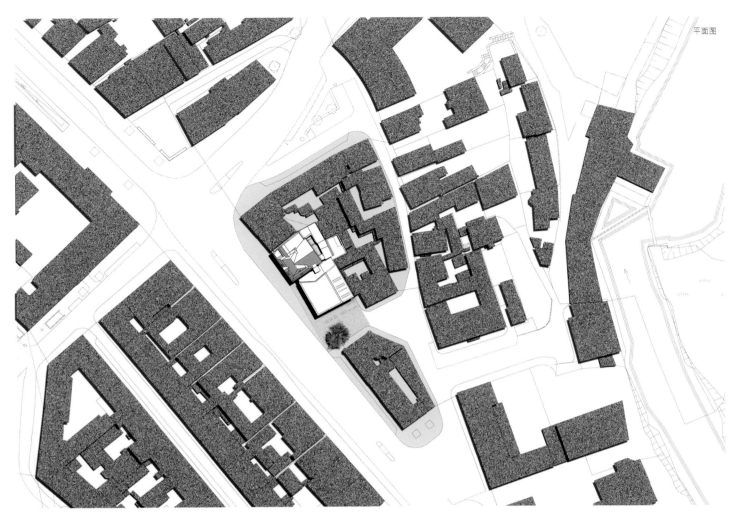

平面图

法兰克尼亚犹太人博物馆

项目地点：法兰克尼亚
设计单位：德国S.I.C.-工程咨询责任有限公司
设计师：杨·古特木特(Jan Gutermuth)、凯南-多马斯
总用地面积：800m²
总建筑面积：1 260m²

GRUNDRISS 1.OG _M:100

GRUNDRISS 2.OG _M:100

　　历史性建筑"法兰克尼亚犹太人博物馆"将通过一个多功能的新建筑进行扩建。此项目为国际竞赛。共有30家设计单位参与了角逐。竞赛评委会在2008年7月25日宣布了获奖情况。共有3家设计单位获得前三名奖项，2家单位获得优秀奖项。S.I.C.的设计方案获得了优秀奖。

　　新的建筑体现了在法兰克尼亚地区的犹太教植根于当地的深厚底蕴及悠久历史。在Fuerth（德国巴伐利亚州的一个城市）的城市景观塑造上采用的建筑材料都取自于当地的砂岩。此建筑与代表法兰克文化的砂岩独块巨石很相似，利用明显的泥沙层堆积的方式进行分隔。新的博物馆建筑与学院博物馆扮演着相似的具有象征意义的角色。

　　显著可见的"泥沙层堆积"在这里象征着通过时间而产生的累积。它让人们认识并明白了弗兰科尼亚地区的犹太人的悠久历史以及他们所产生的影响。它就像地质学家或者考古学家一样，将博物馆的主要功能进行单个的排布，分析以及阐述。从寓意的角度来看，弗兰科尼亚的历史片段将被塑造为一个泥沙层堆积起来的沙石体块，即我们所说的地表岩层，被抬起并站立在城市景观中。

对话：旧—新

　　法兰克尼亚的犹太人博物馆的"最重要的展示"就是它自己具有原本悠久历史的旧建筑。在这个背景下也解释了这栋新建筑体块的形式和方向。

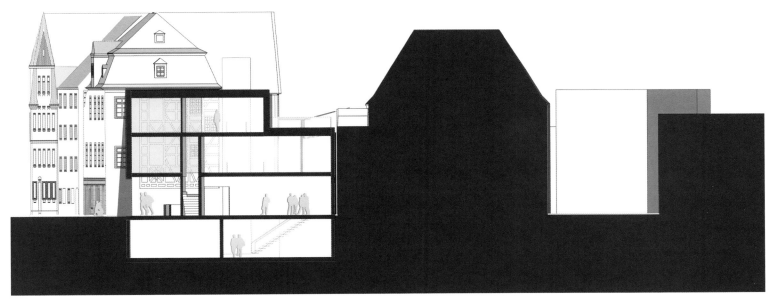
剖面图

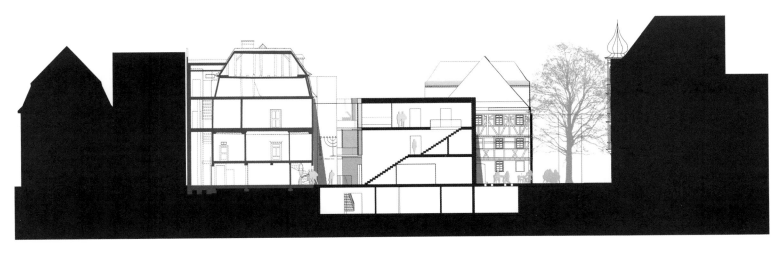
剖面图

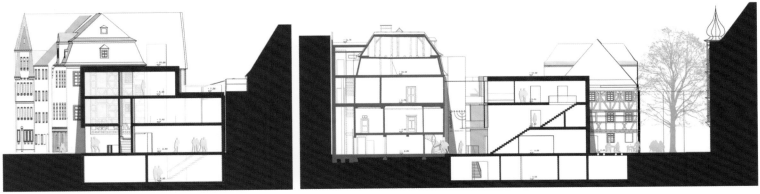
剖面图

MOCAPE-Shenzhen Modern Art Gallery and Urban Planning Exhibition Center

Location: Central Area of Futian District, Shenzhen City
Designed by: German S.I.C.-Engineering Consulting Limited Liability Company
Designer: Stephan Jasper and Jan Gutermuth

Land Area: 29 688.4 m^2
Floor Area: 79 160 m^2
Height: 36m
Floor Area Ratio: 1.8
Building coverage: 44.1%
Greening Ratio: 31%

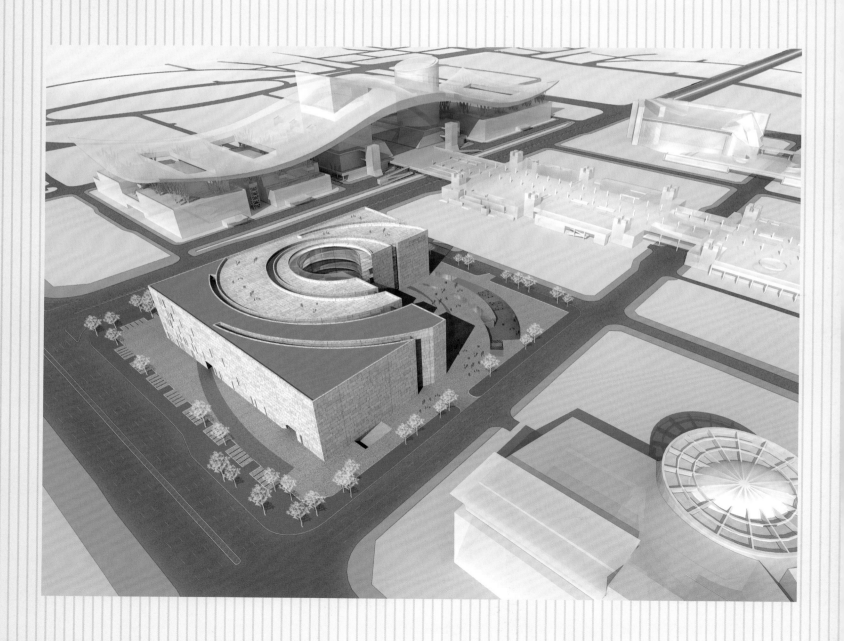

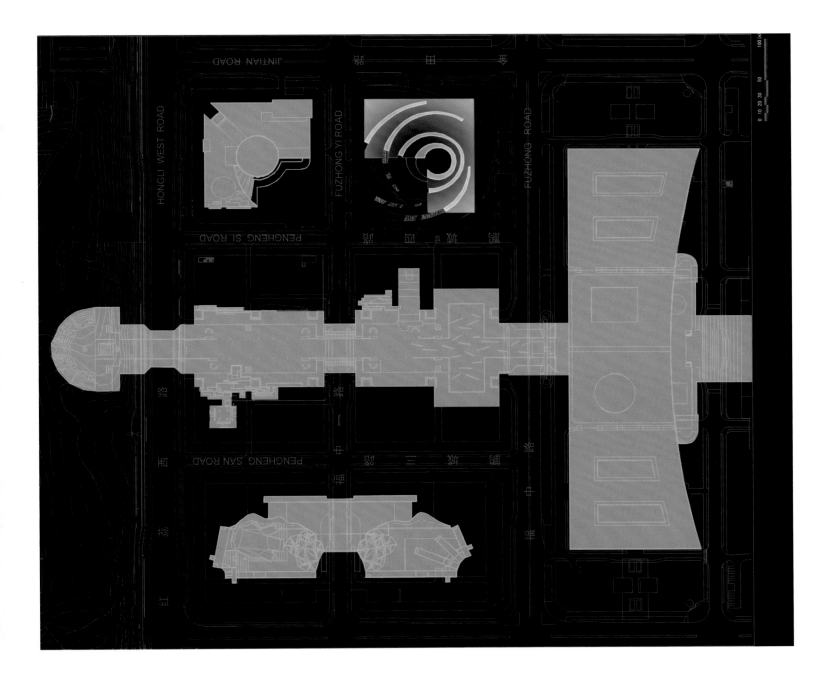

The design treats MOCAPE and children's hall as a whole compared with the integration formed by library and concert hall. Various opposite elements are mixed to form a harmonious and dynamic body. This dialectic approach is the feature of the whole design:

1. Besides this geometric layout, there is simple and generous inside and dynamic "light belt" surrounding the center of construction and guiding sunshine into the chambers of each floor from the clearstory at the roof.

The rhythm of light belt makes a very special internal space and constructs a vivid environmental atmosphere creating the possibility to link and relate different exhibition halls. The light belt has another particularity that it can create an open space and provide the best space for admiration. Meanwhile, it integrates with the functions of construction, links up with internal ramp and stair, and includes traffic system and auxiliary space system into the light belt, and each wide exhibition hall extends between these light belts.

Therefore, there is the possibility of setting various kinds of exhibitions between light belts.

The overall idea is providing an open area in each floor and the area is formed by the "insertion" and "incision" of light belts.

2. Internal space is mostly opened inward with few outward windows and the key point is placed on the two functions, "modern gallery" and "urban planning exhibition center". What forms a strong comparison is the transparent arrangement of entrance and hall.

3. The hall of the first floor and the floors above with the structure of helical light belt are arranged surrounding the hollow cylinder at atrium. Bookstore, exhibition store and other kinds of stores are mainly set in this semi-public area. This kind of arrangement can attract more people who come not just for a visit.

4. Though many surprising visual effects can be created inside the building, the building itself is constructed by the simple frame-shear wall structure to make the cost more economical.

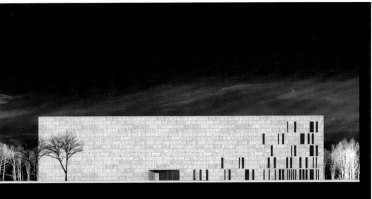

内部交通流线分析图 Interior traffic analysis

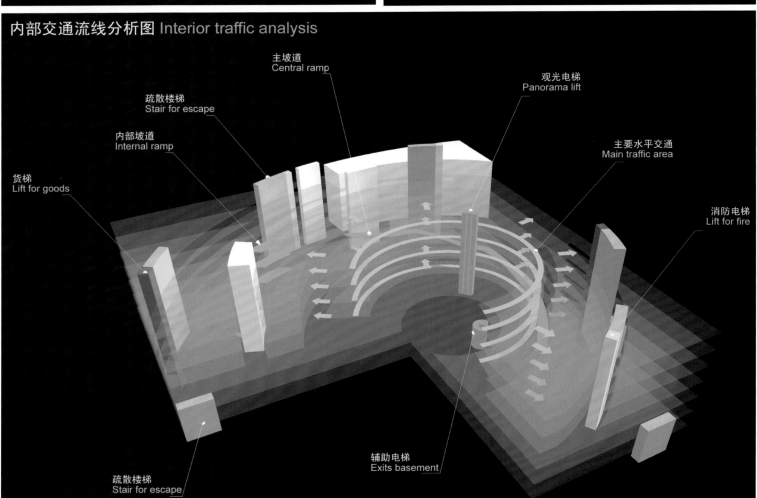

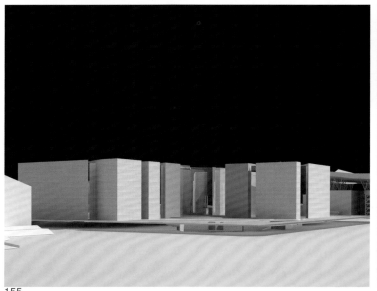
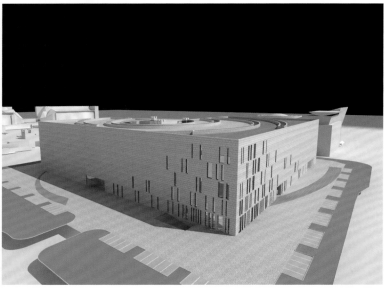

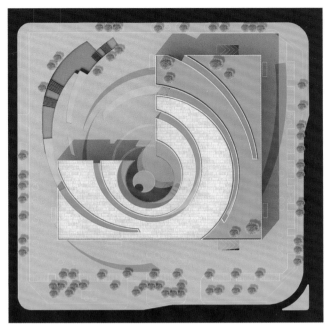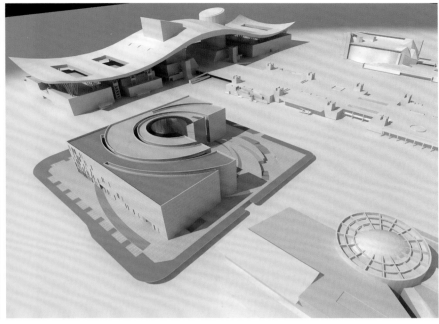

MOCAPE-深圳市当代艺术馆与城市规划展览馆

项目地点：深圳福田中心区
设计：德国S.I.C.-工程咨询责任有限公司
设计师：史蒂芬·亚斯伯（Stephan Jasper）；杨·古特木特(Jan Gutermuth)

用地面积：29 688.4 ㎡
总建筑面积：79 160 ㎡
建筑高度：36 m
容积率：1.8
建筑覆盖率：44.1%
绿化率：31%

设计把MOCAPE与少年宫作为一个整体，与图书馆和音乐厅形成的整体一道审视，将相对立的元素进行融合，并让其形成一个协调的整体，同时充满动感。

这样的辩证手法是整个设计构思的特点：

1.在这个建立在几何布局基础上、简洁大方建筑的内部，设计极具动感的"光带"，围绕建筑中心，从屋顶天窗引阳光进入建筑内部各层内庭。

光带韵律赋予建筑极有特点的内部空间，构建出生动的环境氛围，使不同展厅之间具有连接和联系的可能性。光带还拥有一个特质，即营造开敞空间，为艺术品展示提供最佳的空间观赏环境。同时，它还与建筑功能相结合，与内部坡道和楼梯相衔接，把交通体系及辅助用房体系都集中在光带区内，而各个宽敞的展厅则延伸在这些光带之间。

因此，在介于光带之间的区域则提供了各种类型布展空间的可能。

总体构思，即在每一层都能提供一个开阔的区域，而每个这样的区域都由光带的"插入"和"切割"而形成。

2.内部的空间均向内开口，没有过多对外开口的窗户，将重点放在"当代艺术馆"和"城市规划展览馆"这两个功能上，与其形成强烈对比的则是入口及大厅通透的布置。

3.位于一层的大厅，以及位于其上的含有螺旋光带结构的楼层都围绕着建筑中庭中空的圆柱体切口布置。在这个半公共区域，集中布置书店、展馆商店以及其他各种店铺，这样的布置可吸引到更多的人流，而他们的首要目的可能并不是前来参观展览。

4.建筑的内部空间营造出了很多让人惊奇的视觉效果，但建筑本身的结构是简单的框架剪力墙形式，从而建造成本更为经济。

Guiyang Science & Technology Hall

Location: Guiyang city, Guizhou province
Construction Design: German S.I.C-Engineering Consulting Limited Liability Company
Designer: Jan Gutermuth
Frame: composite structure of reinforced concrete and steel structure

Land Area: 36 630 m²
Floor Area: 22 000 m²
Coverage Ratio: 22%
Floor Area Ratio: 0.6
Greening Ratio: 42.2%
Parking Space: 219

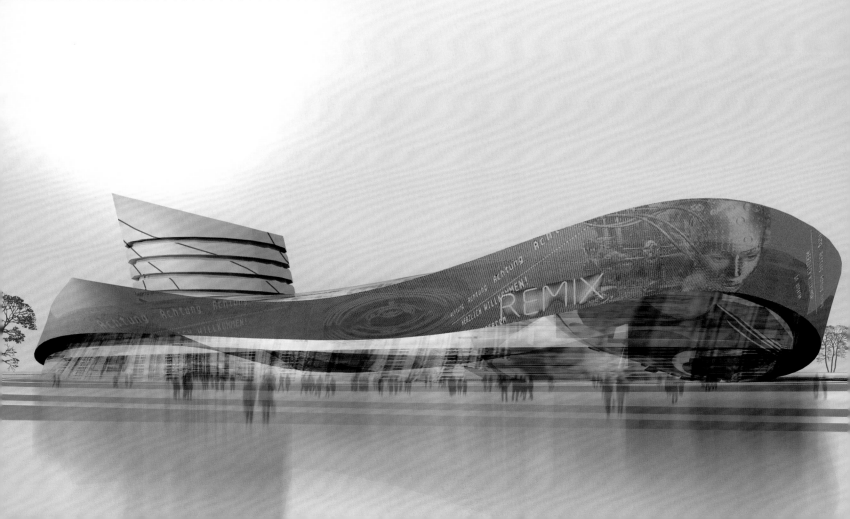

Guiyang Science Museum is symbol of scientific prosperity in the future of Guizhou Province and the sign of economic take-off in western China. It represents innovation and high standards on science and technology and dynamic multi-media applications. S.I.C's design idea for the Science Museum is built on the basis of sign language. The main design element derives from "Mobius Strip". Mobius Strip has only one surface and one edge, in which the unique mathematical property makes it non-orientable. The use of Mobius Strip as the design element for the Museum reflects the mysterious and unknown nature, meaning that people will make never-ending exploration of the nature. The second design element is a distorted cylindrical body. This part has clear geometric form that represents science and the spirit of human to unceasingly explore and study the mysteries of the nature. Solid and stable body firmly grasps the Mobius Strip which is ethereal at the external surface and also extremely precisely tallies and connects with the ground. The entire structure of the building is indicative, while the shape and materials show a sense of innovation.

The whole building sits on the site with an extremely free pattern. Although the main pedestrian and vehicle entrance is located at the northern side of the site, the building itself does not have any clear main facade in traditional sense. The general layout gives a dynamic feel as if an object is in the instant of suddenly going solid in a motion process. The pattern of the building is open and free, creating from all directions an atmosphere of welcome to guide people to the main entrance.

On the landscape, the design also highlights the dynamic character of the building. On the ground around the building, the design creates ring-shaped and streamline paving lines; through greening layout, countless green belts rotate around the building to form a "Tornado" vortex, while the building is like the focal point of the "Tornado". The natural terrain of the site strengthens the dynamic character of the landscape and the whole building, brining the dynamic into a three-dimensional space.

贵阳科技馆（竞赛作品）

项目地点：贵州省贵阳市
建筑设计：德国S.I.C.-工程咨询责任有限公司
设计师：杨·古特木特（Jan Gutermuth）
结构形式：钢筋-混凝土与钢结构的组合结构

基地面积：36 630 ㎡
总建筑面积：22 000 ㎡
覆盖率：22%
容积率：0.6
绿化率：42.2%
停车位：219

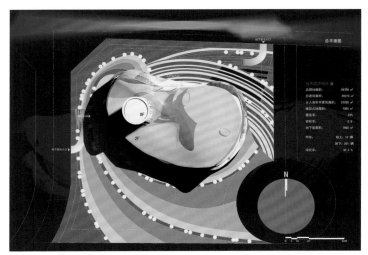

贵阳科学馆是贵州省未来科学技术蓬勃发展的象征，同时也是中国西部经济腾飞的标志。它代表着创新，对科学技术以及多媒体动态应用的高标准。S.I.C.对于科学馆的设计构思建立在符号语言的基础上。主要设计元素来源于"莫比乌斯带（Mobius Strip）"。莫比乌斯带只有一个面和一个边，它独特的数学特性使其具有不可定向性。采用莫比乌斯带作为科学馆的设计元素体现了自然界的神秘以及未知性，寓意着人们将会对自然界进行永无止境的探索。第二个设计元素则是一个被扭曲的圆柱体。这个部分具有清晰的几何外形，它代表着科学，代表着人类对自然界奥秘不断探索和求知的精神。厚实而稳重的体量把飘逸在其外表面的莫比乌斯带紧紧抓住，同时也极其精密地与地面吻合相接。建筑的整个结构具有指示性，而建筑外形与材料则具有创新感。

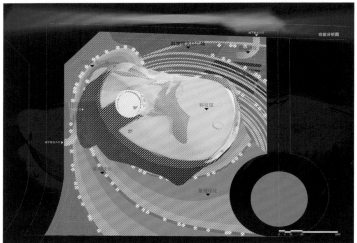

整个建筑以极其自由的形态坐落在场地中，虽然主要人行及车行入口位于场地的北面，但是建筑本身并无明确的、传统概念上的主立面。总平面的规划给人一种动态的感觉，就像是物体在运动的过程中突然凝固的一瞬间。建筑的形态开放而自由，从各个方位都营造出欢迎来访的氛围，并指引人们前往主入口。

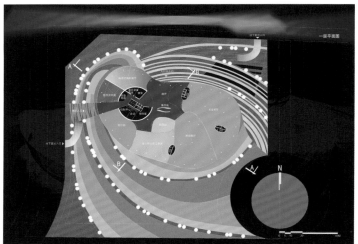

在景观方面，设计也突出了建筑本身的动感特点。在建筑四周的地面，设计设计了环状、流线形的铺地线条，通过绿化布置，无数的绿色带将围绕着建筑旋转，形成"龙卷风"的旋涡，而建筑则好似"龙卷风"的中心点。场地自然的地形则强化了整个建筑及景观的动感特点，它把动感带入了一个三维的空间。

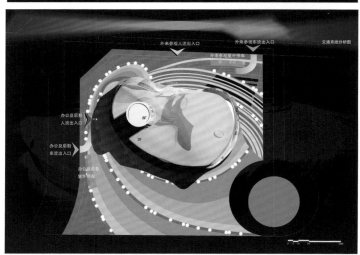

MATRIX
stands for the content of science

2 stories:

Functions:
Exhib. Rooms (12.588m²)
Service Rooms (1.740m²)

material:
transparent, glas construction,

CYLINDER ("fixes" the band)
stands for research, -
 the human pursuit of knowledge

8 stories:

Functions:
Research (1,535m²)
Management/ Logistic (2,047m²)
Scientific Edu.cation (2,047m²)

materials:
solid, steal/ concrete structure

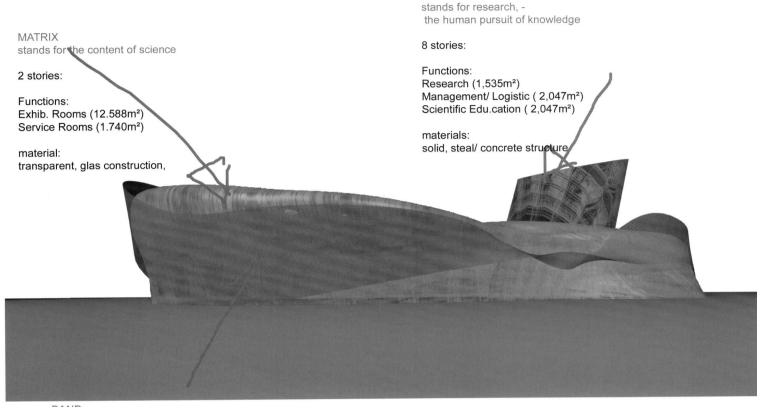

BAND,
stands for nature, science
("MÖBIUS BAND"

material:
Metal gaze - to be projected onto
with film, light etc.

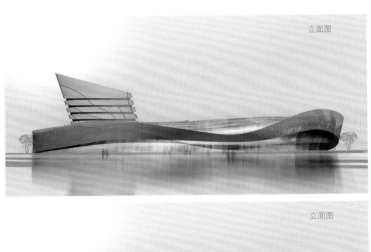
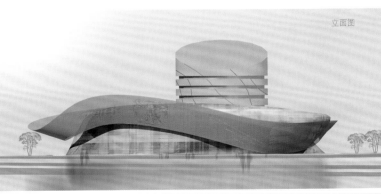
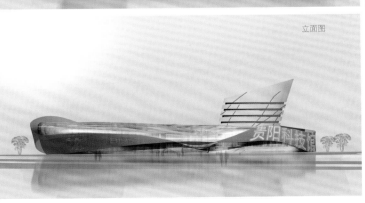
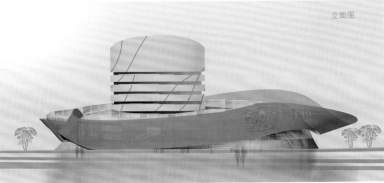

Art gallery of Sichuan Art Institute

Designed by: Dblant Desgin International

Scheme 1:
Land Area: 25800m²
Floor Area: 14727m²
Building Area: 6985m²
Floor Area Ratio: 0.571
Building Density: 27.07%

Characteristic: geometric pattern, min elevation difference
The combination of octagon and square repeats is the recurring subject in the plan. The two perfectly integrate on the plane surface, while they form a traditional barn-shaped body mass with octagon in the dominant position and square in the subordinate position so as to greet to the site which used to be the rolling terraces. The instructive idea of reserving the original ecology as much as possible is carried out during the process of plan to enable Sichuan Fine Arts Institute to remain the rural scenery which appears serene and pristine after numerous buildings are established. Based upon this point, the plan shall be quiet, containing and thought-provoking.

According to the requirements for functions, the building falls into three parts simultaneously with sense of integrity.
The Art Museum for the Fine Arts Institute:

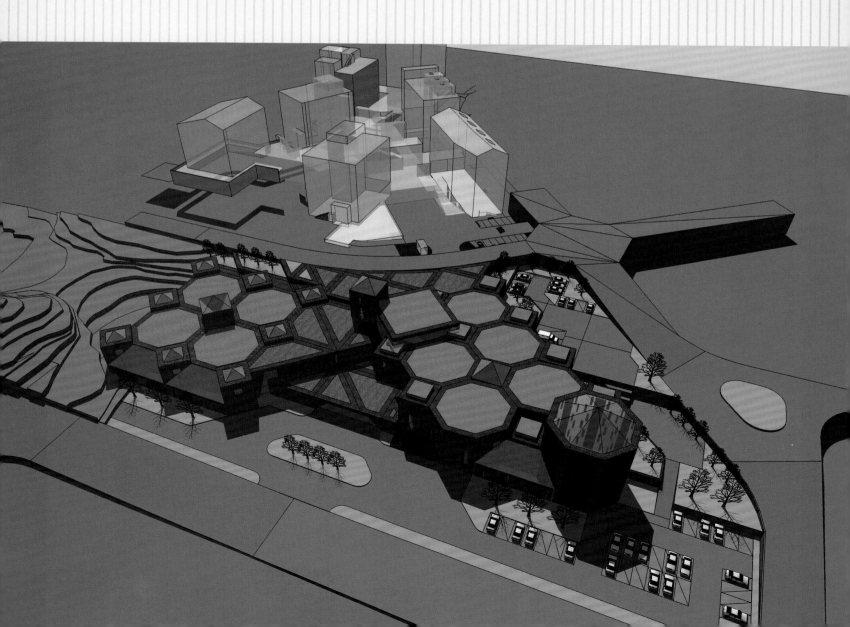

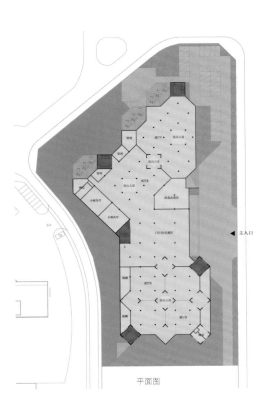
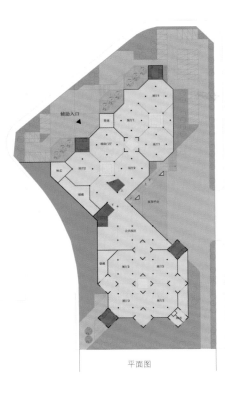
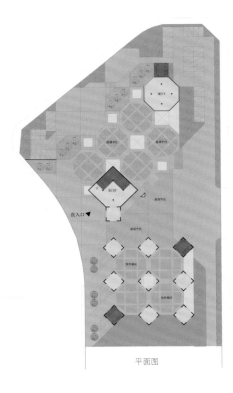

Combined with the public areas, including hall, conference hall, administration office and souvenir store, etc., it is situated at the middle site and forms several squares in different levels in accordance with the terrain and step. It is a relatively open segment.

Master Museum Ⅰ:
It is located in the northern part of the site and stands close to the main entrance of the institute as the unit at the utmost north side, for which reason it is raised and designed to be the "barn" for the purpose of highlighting the building image. The 1st floor above the streamline is connected with the public hall, and there is an auxiliary entrance on the 2nd floor. The roofing of the 2nd floor joined to the west square can be served as the site for outdoor activities.

Master Museum Ⅱ:
It is located in the southern part of the site and it has a complete structure. The geometric surface is in accordance with the barn concept. The 1st floor above the streamline is connected with the public hall, and the vertical traffic rises up to the outdoor exhibition area of the roof.

Scheme 2
Land Area: 25800m^2
Floor Area: 14479m2
Building Area: 6725m^2
Floor Area ratio: 0.561
Building density: 26.06%

Characteristic: sense of sculpture, vague boundary between the building and the site

The starting point of the plan is that what combination form can achieve the maximum uniform for the building and the site. The final debating result is that due to the special functions of the project, the building itself shall be turned into a large-sized art exhibit apart from a container for the work of art. During the design, the building and the site are viewed as a single whole, a sculpture. Floor, wall and roof share the same materials. The part which lowers is the base, while the part which towers is the sculpture.

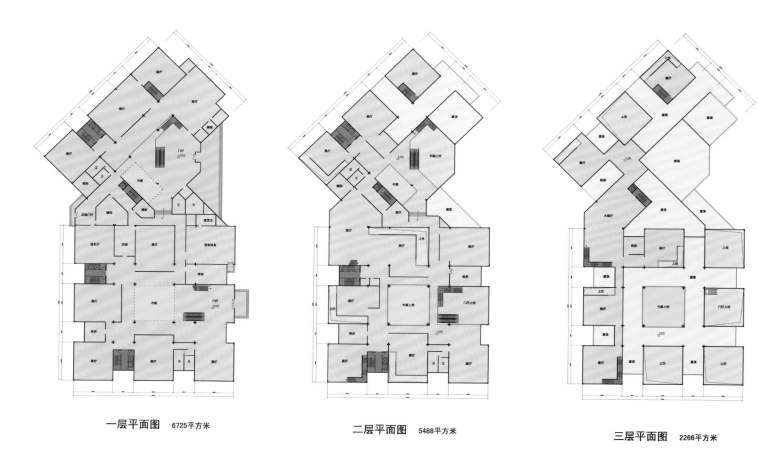

一层平面图　6725平方米　　　　二层平面图　5488平方米　　　　三层平面图　2266平方米

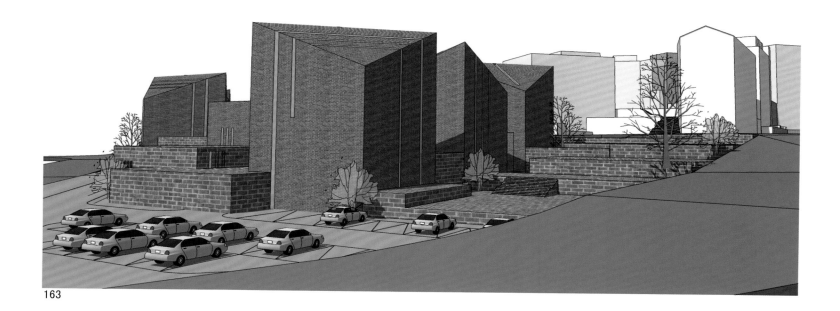

四川美术学院美术馆

建筑设计：都林国际设计

方案一

用地面积：25800m²
建筑面积：14727m²
建筑占地：6985m²
容积率：0.571
建筑密度：27.07%

方案特点：几何图案，最小化高差

八角形与正方形的结合体是方案中反复出现的母题。在平面上，两者完美结合；在立面上，八角形为主，正方形为辅，形成传统的谷仓状体，为的是向这片场地致敬——它曾经是起伏连绵的片片梯田。规划中尽量保留原生态的指导思想，使四川美院在众多建筑落成后，仍然保留着几分宁静质朴的田园风光。基于这一点，方案应该是沉静内敛、耐人寻味的。

按功能要求，建筑分为三大部分，同时具整体感。

美院自用艺术馆：结合公共部分，包括门厅、报告厅、管理办公和纪念品商店等，位居场地中部，跟随地形和台阶在不同标高形成数个广场，是较为开放的部分。

大师馆一：位处场地北部，最北端的一个单元靠近校园大门，故升高并作"谷仓"设计，突出建筑形象。流线上一层与公共门厅相连，二层有辅助入口。二层屋面与西广场相接，可作为室外活动场地。

大师馆二：位处场地南部，形体完整，几何形平面呼应谷仓概念。流线上一层与公共门厅相连，竖向交通上升至屋顶的室外展区。

方案二

用地面积：25800m²
建筑面积：14479m²
建筑占地：6725m²
容 积 率：0.561
建筑密度：26.06%

方案特点：雕塑感，建筑与场地的模糊界限

方案的出发点是用怎样的结合方式能让建筑和场地最大程度统一。最后的讨论结果是因为项目的特殊功能，建筑除了成为艺术品的容器之外，本身也应该成为一座大型的艺术展示品。在设计中，建筑和场地被看做一个整体，一件雕塑，地面、墙身和屋顶使用统一的材料，低伏的部分是基座，高耸的部分是雕塑。

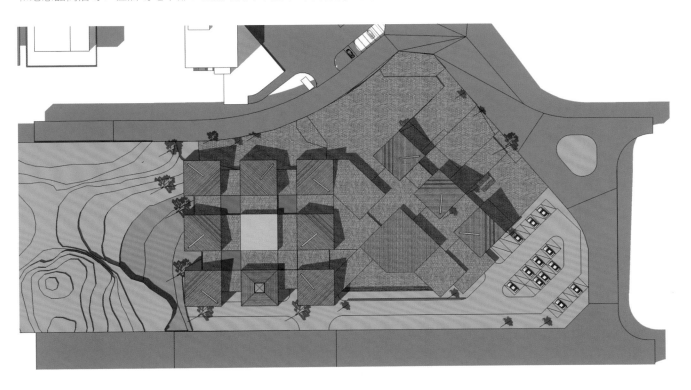

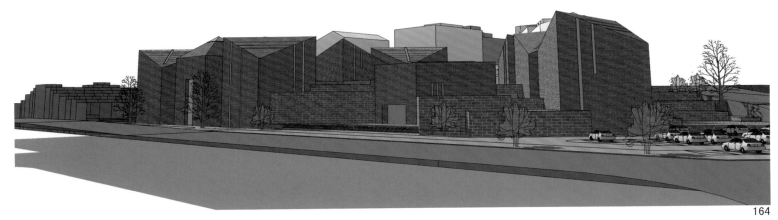

Langfang Grant Theater

Designed by: United Design Group (UDG)

Langfang Great Theater is chosen to be built in Langfang Culture and Art Center and will be built into the core architecture of the Center and the landmark architecture from the basic cultural facilities of Langfang. Langfang Great Theater will be built in the Langfang Urban Culture and Art Center which is a theme park featured by culture and art, occupying 44 ha, with landscape area of 260 thousand square meters, floor area of 82 thousand square meters and investment of 600 million yuan. Stage-I project includes natural cycle park, art park, and literature theme park. Stage-II project contains drama theme park, cultural museum, library and lakeside business district. The Center is composed by 5 landscape areas, including 4 parks which are built by 4 themes, nature, art, drama and literature, namely Natural Cycle Park, Dreaming Lake Park (including 80 000m² water and theatre), Art park, Smart Water Park, together with Road of Wisdom and Calligraphy.

Langfang Culture and Art Center is dedicated to becoming an integrated cultural piazza and art park which is full of historical and cultural connotation and facing future city. It will be one of the most valuable open spaces of Langfang, the important assembly of urban culture and art in northeastern district, an artistic business card of city, the center of social life, a resort for leisure, entertainment and artistic accomplishment, and the "cultural and artistic living room" of local citizens. The Center has become the open space which is full of artistic charm and the place to reflect the charm of city civilization.

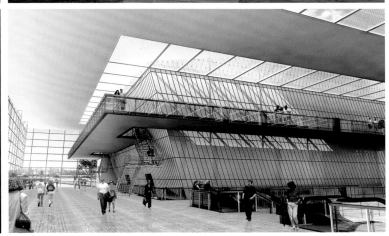
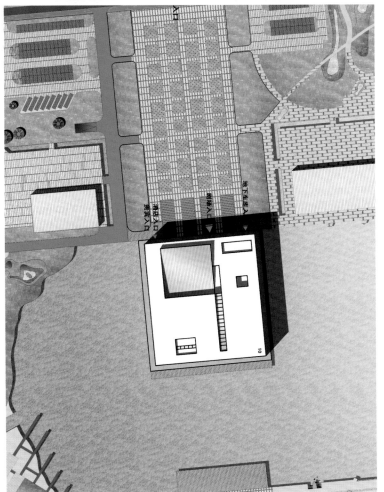

廊坊大剧院

方案设计：UDG联创国际

廊坊大剧院选址位于廊坊市文化艺术中心，将建成为文化艺术中心的核心建筑和廊坊文化基础设施标志性建筑。廊坊大剧院建于廊坊城市文化艺术中心上，廊坊城市文化艺术中心是一个以文化艺术为特色的主题公园，占地面积44ha，景观面积26万m^2，建筑面积8.2万m^2，投资6亿元。一期工程包括自然循环公园、艺术公园、文学主题公园。二期工程包括戏剧主题公园、文博馆、图书馆和滨湖商业街。廊坊城市文化艺术中心以园内的四个公园分别表达自然、艺术、戏剧和文学四个主题，即自然循环公园、梦幻湖（80 000m^2水面+大剧院）公园、艺术公园、益智水上乐园、智慧名言及书法大道五个景区组成。

廊房市文化艺术中心力求创造一个完整的、富有历史文化内涵、面向未来城市的文化广场和艺术公园。它是廊坊市最具价值的开放空间之一，是城市东北区重要的城市文化艺术的集合空间，同时又是城市的一张艺术名片，同也是市民社会生活的中心，为市民提供了一个休闲娱乐、提高艺术修养的好去处，成为当地市民"文化艺术起居室"。文化艺术中心已成为城市中具有公共性、富有艺术感染力又能反映现代都市文明魅力的开放空间。

International Conference and Exhibition City

International Conference and Exhibition City Project are located in the intersection point between Hongqi Street and Haihe Street at Harbin. The entire project is composed of three high-end apartment buildings and two podiums. The main body consists of two plate high-level buildings and a multi-layer building, in which the house's use area ranges from 40 square meters to 110 square meters. The podium integrates 5A office building, comprehensive commercial facility, high-end swimming center and other features. Upon the completion, the project will become a high-grade and trend-leading commercial retail, commercial office, entertainment and fashion lifestyle center. The southeast of the land is planned for a large-scale urban Green Park with great landscape advantage.

The project from planning to architectural design all highlights its landmark feature. In the overall planning, all the street-side architecture show symmetry on both sides of the main entrance to the southeast, forming an axis extending to the Green Park at southeastern corner. The architectural modeling takes axisymmetric approach to form dignified and magnificent main facade, echoing with Victories Hotel. Especially taking into account the landscape effect along the street, the top of architecture is designed with ups and downs and full of changes with a huge dome to form eye-catching identity and enrich the city's skyline contour. The design follows the city's context that the architectural style is modern European style, divided into three sections: the arch windows on the podium, openwork and stencil decorative member and classical detailed molding, not only ensuring the public character of modern commercial office building and the visional permeability but also forming the base in appropriate scale for the main body; at the main entrance, the four-storey Doric portico creates a entrance space with grand momentum; the middle section of the main body is well structured, focusing on depicting the tall and straight impression in vertical direction, combined with large-area glasses to highlight its simple and gleaming modern style; the tall dome at the top and the authentic European classical architectural symbol also highlights the strong and solemn architectural image. The facade design combines both modern and classical styles, using modern materials and classical language and symbols to achieve the sense of force and solidness, creating a new landmark along the Hongqi Street. The design uses tall and straight vertical lines, transparent and bright large windows and multi-level classical symbols as the main facade language, while pays attention to the exquisite detail design, such as the use of multiple sophisticated lines to break the excessive sense of masses, the shutter in the same height on both sides of the main balcony at every two story to increase the elevation level, the progressively back-off lines at the façade corner to enrich the details and so on. The architectural façade uses elegant and bright light yellow anti-stone paint alternated with the tile of the same color for furnishing, forming a unified overall image, luxurious and elegant style and making it stand out from the crowd.

国际会展名城

项目地址：哈尔滨
建筑设计：洲联集团（WWW5A）·五合国际（5+1 Werkhart International）
主设计师：刘力
占地面积：13 000 m²
建筑面积：70 000 m²

国际会展名城项目坐落于哈尔滨红旗大街和海河大街交汇点。项目总该目由三栋高档公寓主楼和两座裙楼组成。主楼为两栋板式高层和一栋多层，户型使用面积从40m²到110m²。裙楼建筑包括5A级写字楼，综合商业设施、高档游泳中心等功能。项目落成后，将成为高档次的、引领潮流的商业零售、商务办公、休闲娱乐及时尚生活中心。用地东南侧为规划城市大型绿化公园，具有绝佳的景观优势。

项目从规划到建筑造型都突出其地标性建筑的特征。在总体规划上，整个沿街部分的建筑沿东南向的主入口两侧呈对称布局，形成了一条延伸至东南角绿化公园的中轴线。建筑造型以中轴对称的方式形成端庄大气的主立面，与华旗饭店遥相呼应。特别考虑到建筑沿街的景观效果，建筑顶部高低起伏，富于变化，以一对巨大的穹顶形成醒目的标识性，丰富城市的天际轮廓线。设计延续城市文脉，建筑风格为现代欧式风格，共分为三段处理：裙楼部分的拱券玻璃窗、镂空花纹装饰构件和古典的细部线脚处理既保证了现代商业办公建筑的公共性及视野的通透性，又形成了主体建筑尺度适宜的基座，主入口处，四层高的陶立克柱廊营造了气势恢宏的入口空间；中段主体建筑体形规整，着重展现垂直方向的挺拔感，配合大面积玻璃，突出自身简洁、剔透的现代风格；顶部的高大穹顶及纯正欧式古典的建筑符号，更突出了雄厚、庄严的建筑形象。立面设计现代与古典并行，利用现代的材料和古典的语言符号实现了建筑的挺拔、厚重感，塑造了红旗大街沿线的新地标。垂直挺拔的竖向线条、通透明亮的大面积玻璃窗和层次丰富的古典符号作为主要的立面语汇，同时注重细节设计的精致完美，如利用多道精致线脚打破过大的体量感、主要阳台两侧每两层设计通高的百叶增加立面层次、在立面转角处设计层层后退的线脚突现丰富细节等。建筑外立面使用明快雅致的淡米黄色防石材涂料与同色面砖相间饰面，整体形象统一，风格典雅华贵，卓尔不群。

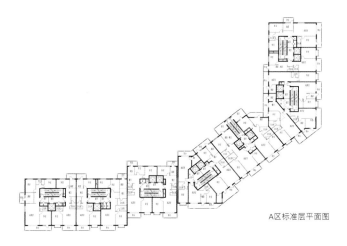
A区标准层平面图

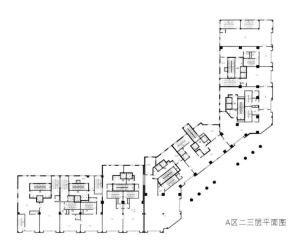
A区二三层平面图

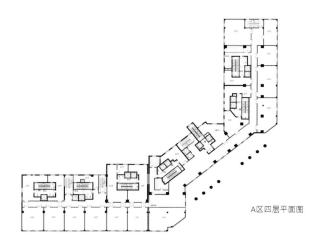
A区四层平面图

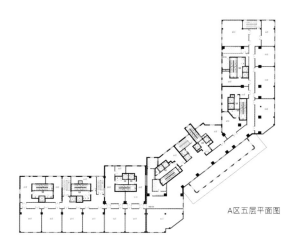
A区五层平面图

A座南立面图

A座东立面图

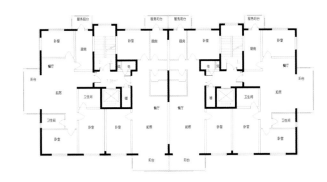
B1标准层平面图

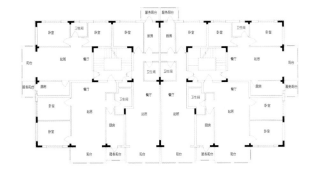
B2标准层平面图

B裙房一层平面　　　　　　　　B裙房二层平面　　　　　　　　B裙房三层平面

B座南立面图

B座东立面图

Shenzhen University City International Conference Center

Location: Nanshan district, Shenzhen city
Designed by: Shenzhen CUBE Architecture Design and Consulting Co., Ltd
Designer: Qiu Huikang, Zhang Zhengqiang, Huang Yanxiang, Zhangyu, Han, Shuyong, Xuechao

Scale: 13265m^2

Shenzhen University Town International Convention Center is sitting in the geographic center of the whole university land, which is also the center of academic exchanges, cultural transmission, the art shows as to the function. The plot is adjacent with the library, in order to give people a pure feeling that the library building is embedded in the mountain mass and consider the original planning and architectural complex, the southern mountain mass has been stretched down, which has formed the slope-shaped earth-sheltered architecture. Most of the construction mass disappears under the natural terrain, coming to be the whole bottom; Entire ring-shaped body rises up from the earth-sheltered part, which

is the visual part of the building. So there is a strong figure ground relationship among the whole complex. Circular point shape is in contrast to the linear-shaped library buildings, echoing harmoniously. Color of buildings is single since the original ones in campus are all in off white hue. The design attaches multiple colors to the risen ring-shaped complex, making it the eye of entire park.

Spatial Analysis:
Blend and penetration between spaces inside and outside: there are two squares separately inside and outside in this architectural layout. The square outside shall surround buildings by half, while builds shall surround all to form the inside square. Therefore the inside square becomes the focus, which is also the outdoor stage, if we see the construction itself to be a big theater. Each space with its function is in scattered layout which is surrounded by ecological natural outdoor spaces.

Three-dimensional, flowing virescence space: the earth-sheltered virescence space on the roof is stretching part of northern mountain mass, while the cortile virescence space is the sinking part of that on the roof. Every function space of the complex is buried by these three-dimensional flowing virescence spaces.

Environment and Square landscape:
The design completes square landscape by means of "puzzle plate", forging it to be a full artistic earth art with the construction and the surrounding natural environment. Roofs and squares are interlaced with each other, sharing everything. Landscape is combining with architectures to create an ecological, comfortable working environment, free and open communication site, expressive life stage together.

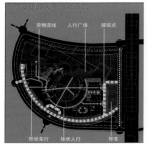

立体开放的内外部人行系统

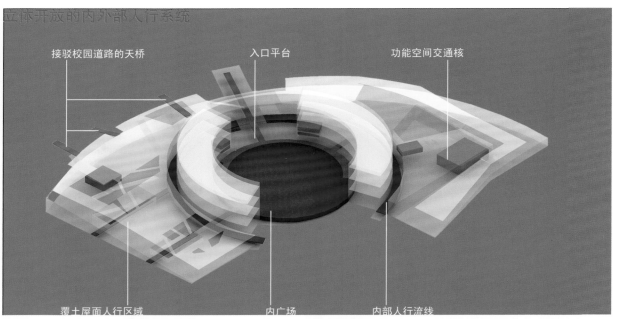

- 接驳校园道路的天桥
- 入口平台
- 功能空间交通核
- 覆土屋面人行区域
- 内广场
- 内部人行流线

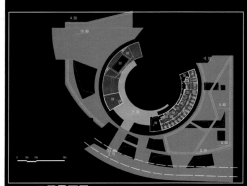

三层平面图

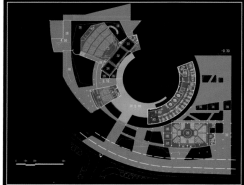

二层平面图

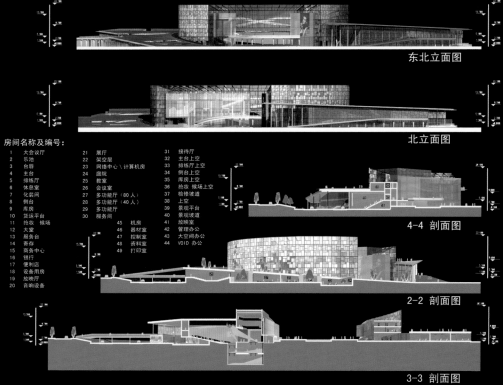

东北立面图

北立面图

4-4 剖面图

2-2 剖面图

3-3 剖面图

房间名称及编号：

1 大会议厅 2 乐池 3 台唇 4 主台 5 排练厅 6 休息室 7 化妆间 8 侧台 9 库房 10 货运平台 11 抢妆 候场 12 大堂 13 服务台 14 寄存 15 商务中心 16 银行 17 便利店 18 设备用房 19 放映厅 20 音响设备 21 展厅 22 架空层 23 网络中心\计算机房 24 庭院 25 教室 26 会议室 27 多功能厅（80人） 28 多功能厅（40人） 29 多功能厅 30 服务间 31 接待厅 32 主台上空 33 排练厅上空 34 侧台上空 35 库房上空 36 抢妆 候场上空 37 检修坡道 38 上空 39 景观平台 40 景观坡道 41 放映室 42 管理办公 43 大空间办公 44 VOID 办公 45 机房 46 器材室 47 控制室 48 资料室 49 打印室

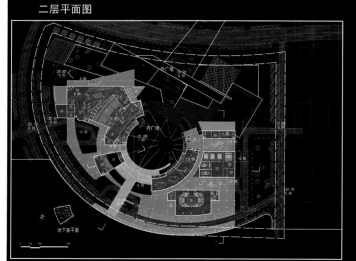

一层平面图

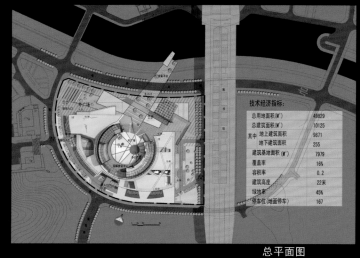

总平面图

技术经济指标：

总用地面积(M²)	49929
总建筑面积(M²)	10125
其中 地上建筑面积	9871
地下建筑面积	255
建筑基地面积	7979
覆盖率	16%
容积率	0.2
建筑高度	22米
绿地率	45%
停车位（地面停车）	167

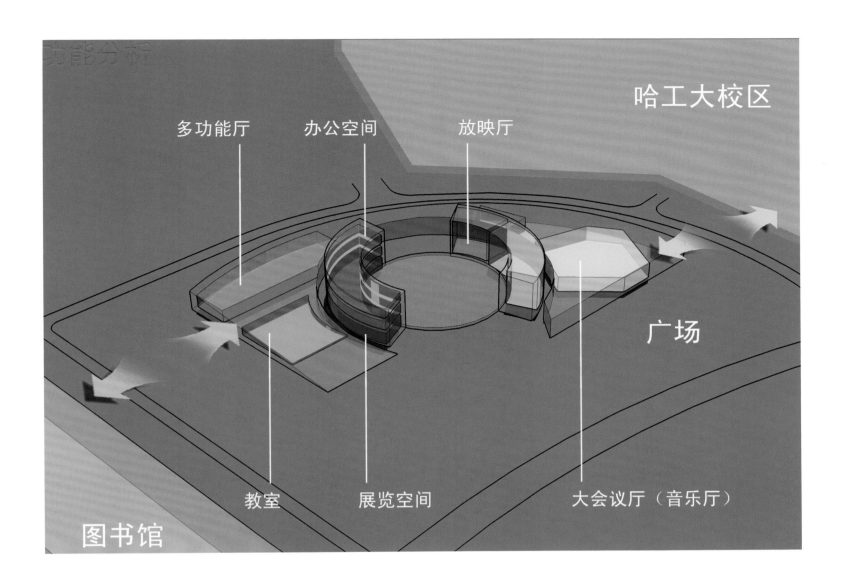

深圳大学城国际会议中心

项目地点：深圳市南山区
建筑设计：深圳立方建筑设计顾问有限公司
设计人员：邱慧康、张政强、黄燕翔、张宇、韩树勇、薛超
项目规模：13 265m²

深圳大学城国际会议中心用地处于整个大学城的地理中心，也是整个大学城学术交流、文化传播、艺术展现的中心。地块同图书馆相邻，为了尊重原规划和已建建筑,保证图书馆建筑嵌入到山体的纯粹感觉，将地块南边的山体延续下来，地块上形成坡状的覆土建筑。大部分建筑体块消失在自然地形之下，形成整个建筑底；完形的环状体量从覆土的部分升起来，是整个建筑的可视部分，整个建筑形成强烈的图底关系。圆形的点状同图书馆建筑的线状形成对比，和谐呼应。校园现有建筑多采用灰白色调,因此建筑色彩单一。项目处于园区的中心位置，设计给升起的圆环附上色彩斑斓的颜色，使之成为整个园区的点睛之笔。

空间分析：

室内外空间的交融、渗透：建筑空间布局上，形成内外二个广场，外广场半包围着建筑，而建筑围合出内广场，内广场是建筑的焦点，是建筑的户外舞台，建筑本身是个大的剧场。建筑的各功能空间分散布局，生态自然的室外空间包围着各种功能的室内空间，

立体、流动的绿化空间：覆土的屋顶绿化空间是地块北边山体的延续，而建筑的中庭绿化空间则是屋顶绿化空间的下沉，建筑的各功能空间就是被这些立体的流动的绿化空间所掩埋。

环境及广场景观：

设计运用"拼图板"的方式完成广场景观，使之同建筑及地块周围的自然环境形成一幅完整丰富的地景艺术。屋顶和广场相互交织着，不分彼此。景观同建筑结合，一起创造出生态、惬意的工作环境、自由开放的交流场所、乐于表现的人生舞台。

Suzhou Conference Center

Planed & Designed by: Beijing Oriental Huatai Architectural Design Project Company Ltd.

There is no need to build a plane central structure in Taihu Lake Resort which is the new image representative of Suzhou. How to coordinate the modern and classical elements into the structure and plots, even make it outstanding? The design starts with the Suzhou Taihu Lake culture, the Suzhou garden culture, Suzhou gardening culture and Suzhou modern culture to conclude the modern body and space which are full of ancient culture image metaphor. The perfect modern buildings with characteristic of restraining are standing in Taihu Lake region, becoming the new landmark of Suzhou. Laurel is Suzhou's representative flower, which is applied into the pattern of printing glass curtain walls. The semi-white curtain wall is just like a white sail standing on the bank, while abundant colors inside are pinching through the walls with sense of silk, which seems like broidery, attacking people who come here. Little holes in the windows in Taihu Lake Stones openwork pattern makes the whole building look like an expanded miniascape, standing in the water.

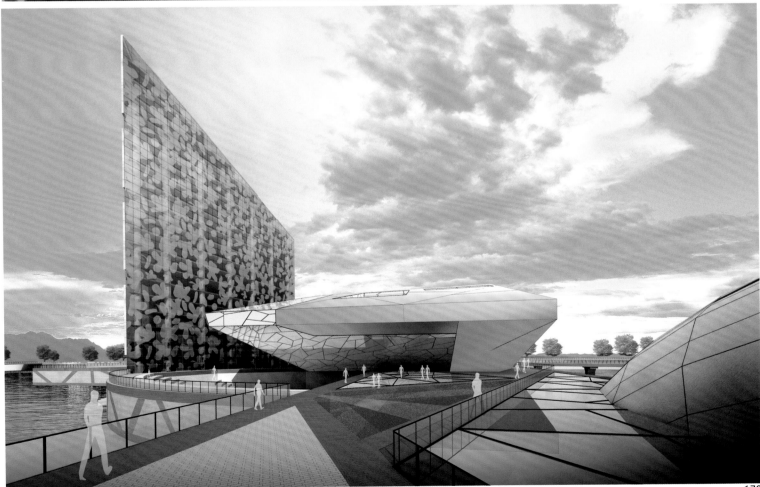

178

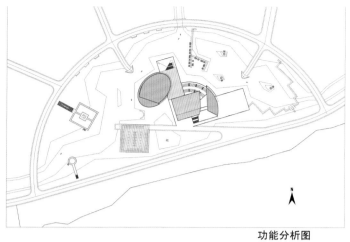

功能分析图 　　　　　　　　　　　　　　　　　　　交通分析图

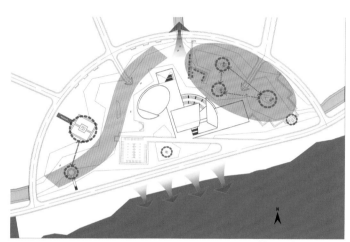

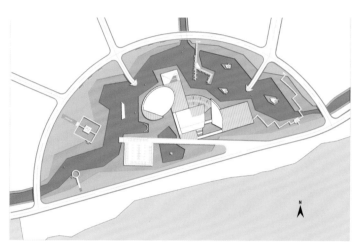

景观分析图 　　　　　　　　　　　　　　　　　　　绿化分析图

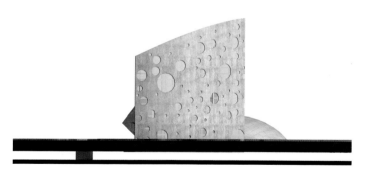

东立面 　　　　　　　　　　　　　　　　　　　　　南立面

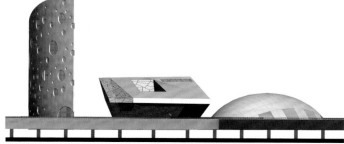

西立面 　　　　　　　　　　　　　　　　　　　　　北立面

苏州会议中心

规划/建筑设计：北京东方华太建筑设计工程有限责任公司

代表苏州新形象的太湖度假区不需要一个平庸的中心建筑，如何把现代和古典融入建筑与场地中并使其脱颖而出？设计从苏州太湖文化、苏州园林文化、苏州园艺文化、苏州现代文化的研究出发，得出一些充满隐喻的指向古老文化形象的现代形体和空间，完美的具有内敛精神特质的现代建筑立于太湖之中，成为苏州的新标志。桂花是苏州的市花，设计将它用于印花玻璃幕墙的图案，半透明的白色幕墙像一张白帆立于湖畔，而室内丰富的色彩透过具有丝绸质感的墙面，正如刺绣一般吸引住每一个来到这里的人。如太湖石镂空形式的窗洞使得整个大楼如一个放大的盆景伫立水中。

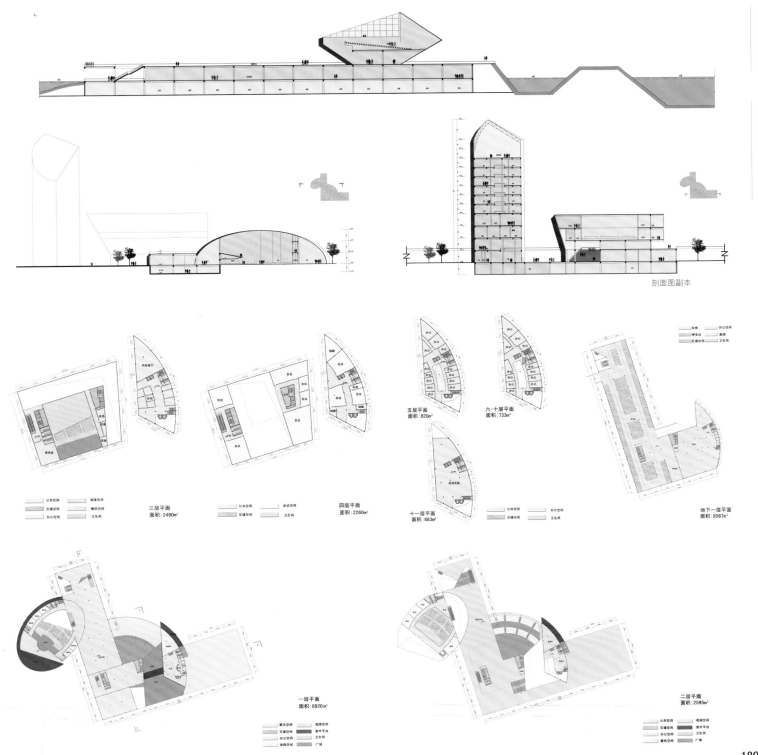

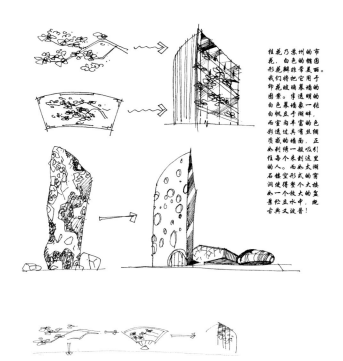 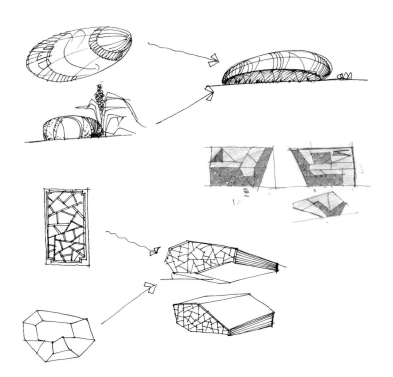

桂花乃苏州的市花，色泽鲜艳，白色非常漂亮。我们将把印花玻璃幕墙用于整个建筑的外立面，而室内透过彩色玻璃看外面的景色如同一幅美丽的图案。幕墙立面的丝网印花正如一张纱，吸引着每一个来到太湖边的人。而窗形式的石楼空洞使得整个太湖楼盘兀立水中，如一个敦伦古典又波普的雕塑！

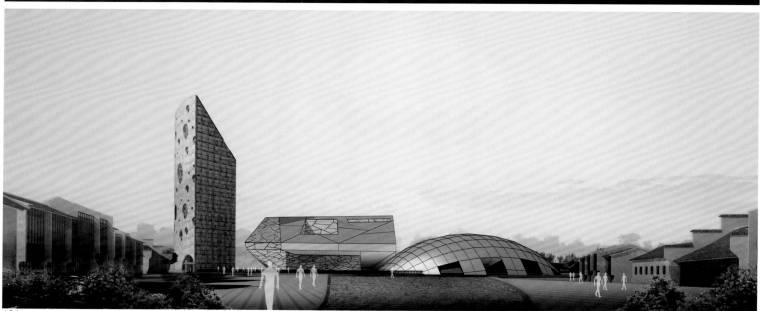

中国太湖文化论坛会议中心 China Tai Lake Culture Forum Conference Centre

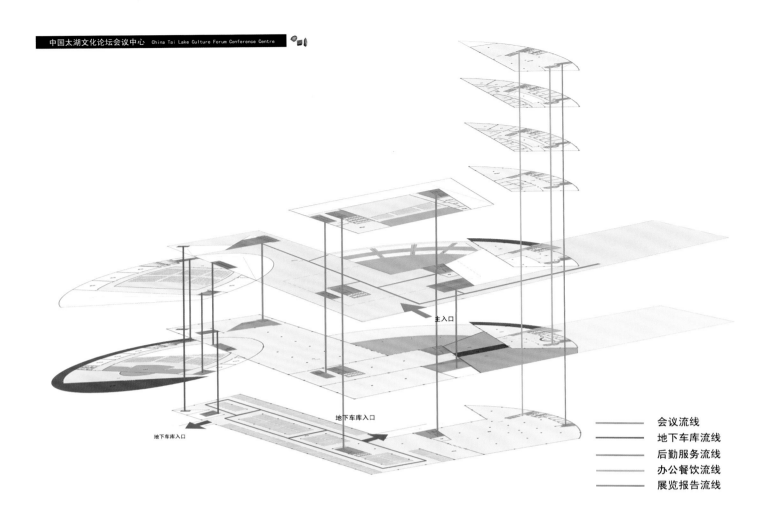

主入口
地下车库入口
地下车库入口

—— 会议流线
—— 地下车库流线
—— 后勤服务流线
—— 办公餐饮流线
—— 展览报告流线

Layout of Hutang technology center

Designed by Dblant Desgin International
Land Area: 235.93 ha
Floor Area: 133.47 ha

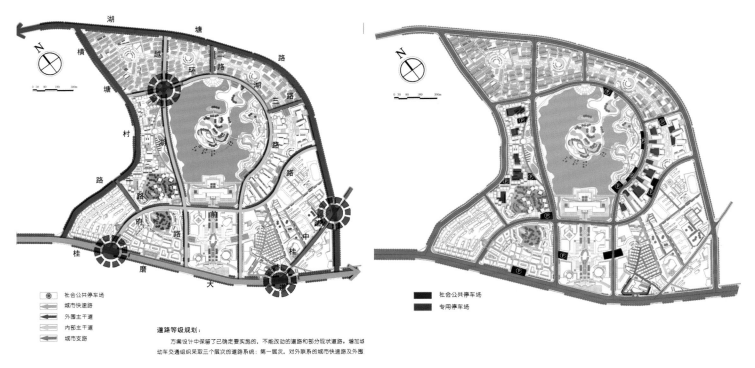

Layout of Hutang technology center in new town of eastern Lijiang Guilin takes full advantage of the river way landscape to create a waterfront leisure space which is full of intimacy and a distinctive, modern urban center. Main function of the plot is for Open Park and an integrated waterfront business district. Central five-star hotel is a landscape architecture with clear features, high visibility and strong mark. Buildings in business center stand around the central island in a gesture of dropping into the water, novel modeling with flavor of the times. Revetment is in the form of hydrophilic or soft shorelines, there are back-line districts, leisure and green belts, bicycle zone, landscape green belt, riverside walking area in rich and varied form.

桂林漓东科技新城湖塘科技中心

规划设计：都林国际设计
规划总用地：235.93ha
总建筑面积：133.47ha

桂林漓东科技新城湖塘科技中心规划设计充分利用河道的景观作用，营造一个充满亲切感的滨水休闲空间，同时创造一个富有特色的现代城市中心。地块功能主要为开放性公园、滨水综合商业区。中心五星级酒店为景观建筑，特征鲜明、形象突出，标志性强。商业中心建筑环绕中心岛并向水跌落，造型新颖，具有时代气息。驳岸形式为亲水岸线及软质岸线，有建筑退线带、休闲绿化带、自行车带、景观绿化带、滨水漫步带等形式丰富多样。

New Embassy of the Republic of Turkey in Berlin, Germany

Location: Berlin, Germany
Designed by: German S.I.C-Engineering and Consulting Limited Liabilities Company
Designer: Jan Gutermuth

Base Area: 5 000 m²
Floor Area: 8 000 m²

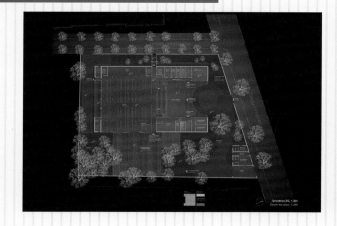

The history of Turkey is with rich oriental flavor and characteristic. It not only connects the east and west in geography and is always the focus of the communication and intercourse between the east and west. If we are to embody the status of Turkey in the history through the art of architecture, the words "confidence", "mediation" and "intermediary" can be used to express the uniqueness of this country. The enclosure of two wings forms a space with a transparent atrium inside and also shapes the gestures of "Welcome" and "Self-Protection".

The theme "The communication and intercourse between the east and west" is also shown in the design of facade. Facade integrating modern elements and traditional elements is related with the typical and western facade to form a comparison.

The architecture opens to the city with a kind of proper, representative and attracting way.

立面图

立面图

立面图

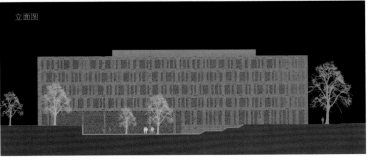
立面图

剖面图

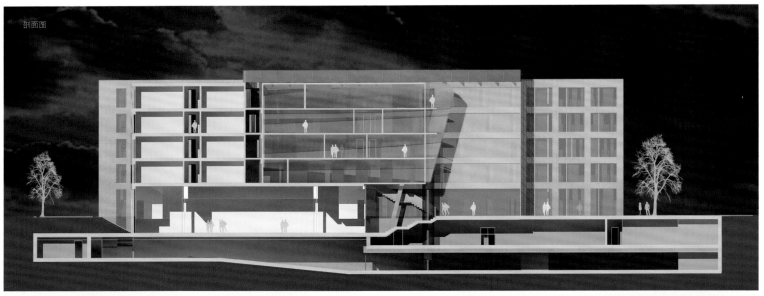

剖面图

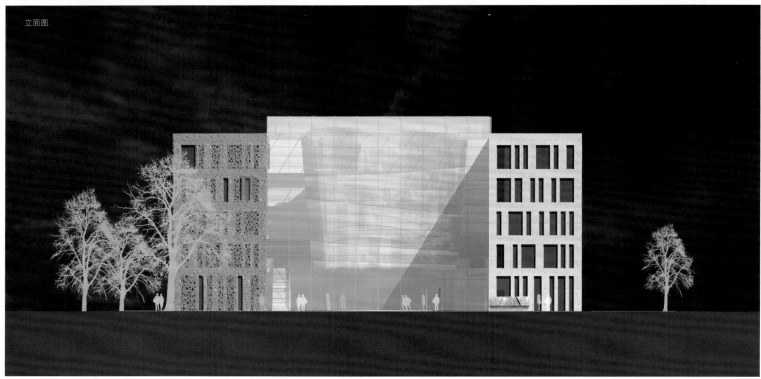

立面图

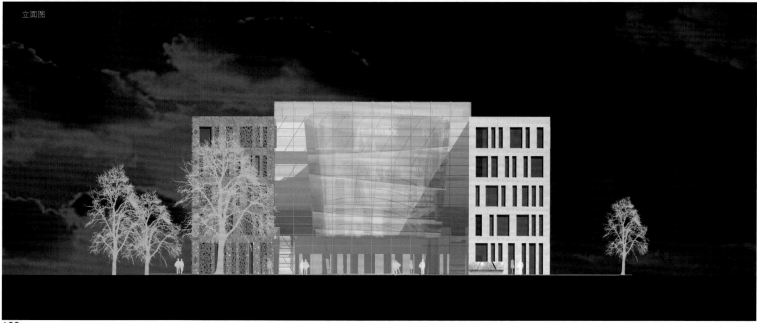

立面图

土耳其共和国驻德国柏林大使馆新馆

项目地点：德国柏林

设计：德国S.I.C.-工程咨询责任有限公司

设计师：杨·古特木特(Jan Gutermuth)

基地面积：5 000 m²

总建筑面积：8 000 m²

土耳其的历史具有丰富的东方色彩和特色，它不仅在地理上连接着东方与西方，而且一直是东西方间的对话与交流的焦点所在。如果把土耳其在历史上逐渐形成的地位通过建筑艺术进行体现，那么"自信"、"中间调停"及"中间媒介"则是表达这个国家特质的方式。

建筑的两翼相围合所形成空间构造了一个通透的中庭，同时也营造出"欢迎"和"自我保护"相并存的姿态。

"东西方对话与交流"这个中心主题同样也体现在建筑立面的设计中：集现代元素和土耳其传统元素为一体的立面与典型的西方特色的立面相互联系，并形成对比。

建筑以一种恰当的、具有代表性的且极富吸引力的方式向城市开放。

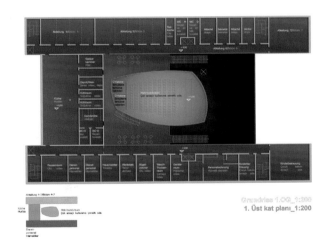

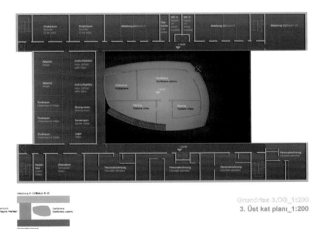

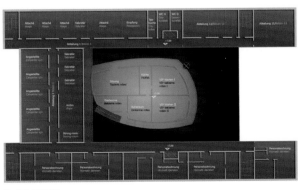

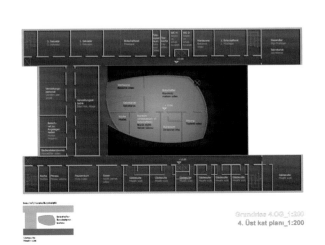

Guiyang Longdongbu International Airport terminal building

Location: Guiyang
Scheme Design: UDG United Design International

Land Area: 35 800m²
Floor Area: 108 210m²

The designed target for Phase II Terminal Expansion of Guiyang Longdongbu International Airport is by the year 2020 to achieve the annual passenger throughput of 15.5 million persons, cargo throughput of 220,000 tons and 150,000 aircraft sorties. The main construction contents are at the north side of the existing terminal (Zhaojiawan Lot) to build the second terminal with the total area of 110,000 square meters; on both northern and southern sides of the existing terminal to build a new apron covering 258,000 square meters so that the total parking bays reach 50, while to build a new cargo warehouse of 19,500 square meters, a new special garage of 6,000 square meters, as well as supporting fire and rescue, power supply, water supply, oil supply and other production and living facilities.

Phase II Terminal Construction Program requires the space form to have a rich sense of times, so the shape seeks to be magnificent and simple with ethnic and geographical characteristics; the space layout is characterized with coordination to make full use of the old terminal, focusing on new building to make smooth transition from the old to the new. The modular combination is the first choice; the rationality in the short-term and long-term construction: room for time or space should be left and the upcoming and future terminals should be placed single-sided as much as possible (both on the west side of the runway); functional practicality: people-oriented and efficient for passenger convenience and comfort. Air side and land side should be compact and reasonable with convenient traffic flow line. The near passenger bridges are optimally equipped in terms of quantity; structural optimization is the key to project cost economy; the advanced science and technology puts emphasis on green concept, saving energy, water and land, protecting the environment and making full use of the terrain.

The morphological theme of the terminal program design is "flow." The flowing form coincides with the rationale in nature and life to inspire our positive and progressive associations.

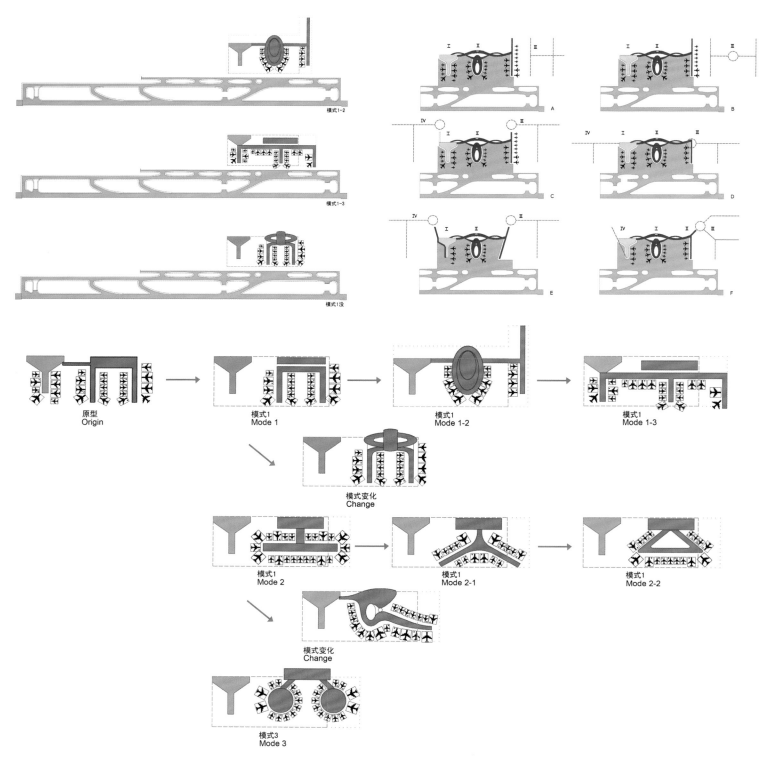

贵阳龙洞堡国际机场航站楼

项目地址：贵阳
方案设计：UDG联创国际

用地面积：35 800m²
总建筑面积：108 210m²

贵阳龙洞堡国际机场二期航站楼扩建设计目标年为2020年，预计年旅客吞吐量1550万人次、货邮吞吐量22万吨、飞机起降15万架次。主要建设内容为：在现有航站楼的北侧(赵家湾地段)新建总面积11万平方米的第二航站楼，在现有航站楼南北两侧新建25.8万平方米的停机坪，使机场总停机位达到近50个，同时新建1.95万平方米的货运仓库，新建6000平方米的特种车库，以及配套消防救援、供电、供水、供油等生产、生活辅助设施。

二期航站楼建设方案要求空间形态富有时代感，造型力求大气、简洁，有民族、地域文化特色；空间布局具有协调性，充分利用好老航站楼，以新楼为主，新、老楼过渡流畅。以单元式组合为首选。近远期建设的合理性：时间、空间方面留有余地，近远期航站楼尽可能单面（均在跑道西侧）布置。功能的实用性：以人为本，高效，旅客方便、舒适。空侧、陆侧紧凑合理，交通流线便捷。近机位数量配置最优。结构最优化是项目造价经济性的关键。强调科技的先进性以及绿色理念。节能、节水、节地、环保，充分利用地形。

该航站楼方案的设计形态学主题是"流动"。流动的形态暗合自然，生命的至理，激发人们积极、向上的联想。

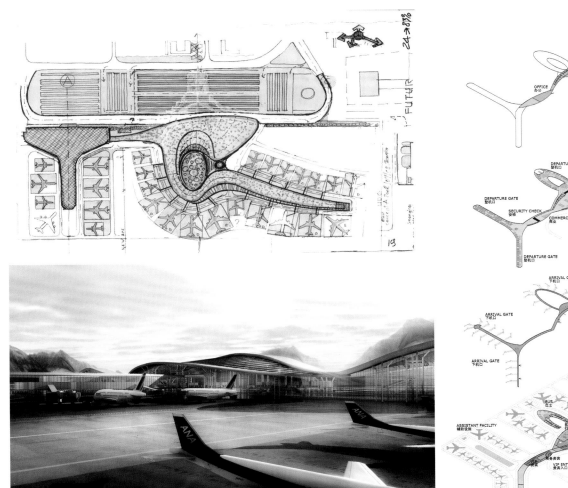
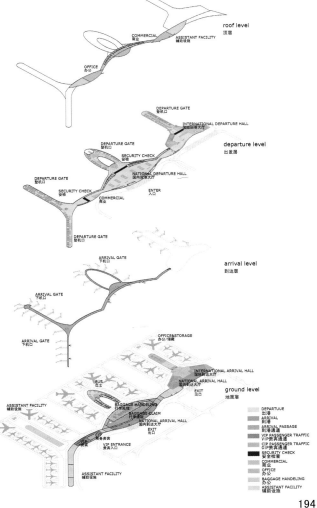

HOSPITAL & CULTURE
ARCHITECTURE

医疗文化建筑

Changzhou Xinbei People's Hospital

Location: Changzhou city, Jiangsu province
Planning/Architecture Design: Zhejiang Modern Architecture Design Institute Co., Ltd
Scheme Design: Lichen, Xie Zuochan, Jinlei, Liuliu, Xujun
Floor Area: 128 000 m^2

The project is set in the south of Xinbei development zone of Changzhou city and surrounded by roads with Changjiang Road which is the main trunk road to the east and an academy to the west. The land is flat with convenient traffic and ideal for medical buildings. The hospital will be a modern and comprehensive building cluster integrating medical service, research, health care and disease prevention into one offering 1200 beds. The land looks a trapezia with the distance from south to north longer than that from east to

west which is defined by a curved road. With ellipse as the interface of annex, designers intend to let people seek for appropriate hospital texture between freedom and restriction and construct a unique space in the area. The Ellipse interface can not only serve for an open and directive entrance but also make a vivid facade on Changjiang Road. Comparison and coordination are valued in the layout, for example, the elliptic lecture hall in the northwestern corner correlates with the main body of annex; clinic in the southwestern corner and the living area in northeastern corner. The middle adjoins with the two impatient buildings to create a uniform and changeful space configuration. In the layout, the construction consists of two clusters located in the south and north. They will not only reasonably meet the demands of functions and also shape a diverse, ruly and flexible space.

Medical comprehensive building is composed by curve body and two high-rise impatient buildings and elliptic roof extends to the impatient buildings. Designers use the curved annex and square main building to form a kind of occlusion connection and construct a lively and generous space. The oneness and integrality of hospital is expressed by simple composition technique. Sequence assembly of unit module is adopted to form the coordinative relation and heterogeneous elements are inserted into the large facade to activate the space configuration. The design highlights the vertical line of main building to offer a rising trend; annex is flatly processed to form an unfolded facade. The designers always seize on showing the generous image of hospital and adopt the concise but not simple design technique to exhibit the exquisite and harmonious spatial image of Changzhou Xinbei People's Hospital.

常州市新北人民医院

项目地址：江苏省常州市
规划/建筑设计：浙江省现代建筑设计研究院有限公司
方案设计：李晨、谢作产、金磊、刘浏、徐俊
总建筑面积：128 000m²

常州市新北人民医院项目位于常州市新北开发区南侧，四面环路，东侧临主干道长江路，西侧为专科院校。用地平坦，交通便捷，是一处较为理想的医疗建设用地。新建医院共设1200床，将是集医疗、科研、保健预防为一体的现代化综合性医疗建筑群体。用地呈梯形，南北长、东西短，西边以弧形道路为界。本案裙房采用椭圆形界面，在自由与约束中寻求合适的医院肌理，形成该区域特有的空间形态。椭圆形界面既可以形成入口广场的开放性及入口导向性，又能在长江路上产生活跃的沿街立面。规划注重形体的对照与呼应，西北角设计椭圆形报告厅与主体裙房呼应，西南角的门诊体块与东北角的生活区体块相呼应，中间与二栋住院楼衔接，产生统一而富于变化的空间构图，规划中建筑形成南北两大组团，合理的满足功能要求，同时形成错落有致、收放自如的建筑空间。

医疗综合楼由弧形体块与二栋高层住院楼组合成建筑群体，椭圆形屋顶板延伸至住院楼，建筑利用弧形裙房与方形主楼产生咬合关系，构筑活跃而大气的空间氛围。运用简约的构图手法，表达医院空间的统一性及完整性。设计中采用单元模块的序列组合，形成呼应关系。在统一的大面上穿插异质元素，活跃空间构图。主楼强调竖向线条，产生升腾动势；裙房作水平处理，形成舒展立面。设计中始终把握舒展大气的医院总体形象，采用简约而不简单的造型设计手法，展现常州市新北人民医院精致和谐的空间形象。

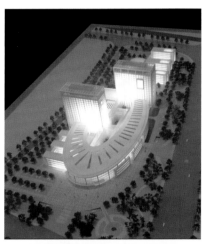
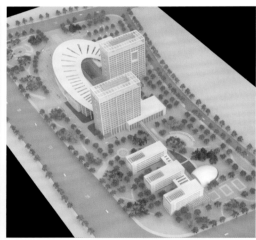
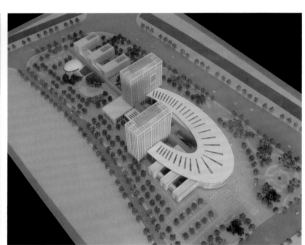

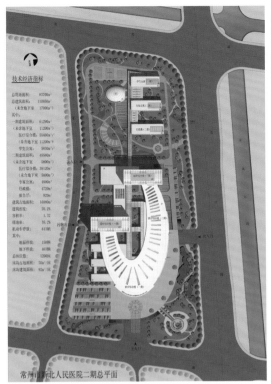
常州市新北人民医院二期总平面

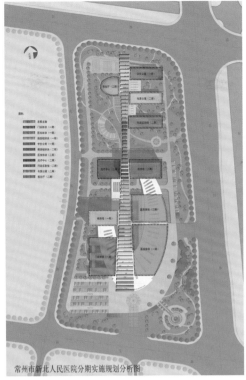
常州市新北人民医院分期实施规划分析图

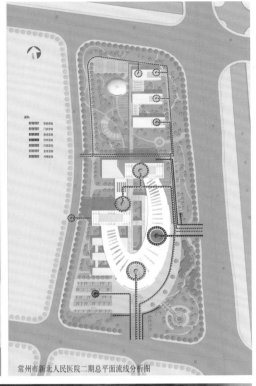
常州市新北人民医院二期总平面流线分析图

Hangzhou Xiasha Hospital

Location: North of Genshan East Road, Xiasha Town, Economic and Technical Development Zone, Hangzhou City
Planned and Designed: Zhejiang Modern Architectural Design & Research Institute Co., Ltd
Designer: Li Kejun, Xutao, Feng Sufen
Floor Area: 170 000 m^2
Storey: 22
Frame: frame-shear wall structure

Hangzhou Xiasha Hospital is located in the north of Genshan East Road, Xiasha Town, Hangzhou Economic and Technical Development Zone, close to Xingfu River and Xingfu Road to the east and Qige North Road to the west; Nanyuan Road divides the land into northern and southern regions. The location has convenient traffic, flat terrain and elegant environment. The hospital will offer 1200 beds, with 22 stories on the ground and 1 storey underground, which includes emergency department, medical technology department, inpatient department, logistics department, administrative department, R&D department and healthcare department, etc.

This scheme fully develops the properties of terrain and outstands the uniqueness and difference of construction and emphasizes the comparison and alternation of construction body; plane and facade both show the texture of sculpture to so as to create a unique and rhythmic construction personality; designers use liquid-ink-style tone----abundant gray series to be the main color of architecture; introduce water body to interact and combine with the building and form the image of a modern architecture with vivid characteristics and rich spatial shapes. Starting from symbolism and regional features, the scheme is to build a modern medical construction. The integral shape symbolizes a beating heart and embodies the sacred responsibility of a hospital to continue the life.

The scheme divides the land into two functional sections with Nanyuan Road as demarcation line, the northern and southern sections. Southern section is designed to offer medical treatment and northern section is built for offering logistic services for administrative and scientific research. Medical section equipped with medical complex including clinic, emergency, medical technology and sickroom. The land is shaped like a rectangle, with eastern and western sides wide and southern and northern sides narrow, so

that the design arranges the medical main stem transversely, which connects each functional section together, and medical complex is placed to the western side of land as close as possible, so as to reserve the space for main entrance hall in the south, central greenbelt in the east and future development. Administrative and scientific research logistic section is equipped with administrative and research building and health industrial section. Inside the building there is administrative office, teaching and research center, kitchen, dining-room and duty room which are closely related with medial complex through the routeway on the ground and underground.

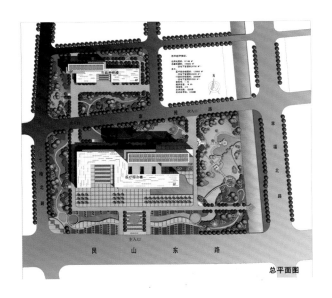

总平面图

杭州市下沙医院

项目地址：杭州市经济技术开发区下沙镇艮山东路以北
规划/建筑设计：浙江省现代建筑设计研究院有限公司
建筑师：李科军、徐涛、封素芬
建筑面积：170 000 m²
建筑层数：22
建筑结构：框架剪力墙

杭州市下沙医院位于杭州市经济技术开发区下沙镇艮山东路以北，东部紧邻幸福河及幸福路，西邻七格北路，南苑路将地块分为南、北两区。该地块交通便捷、地势平坦、环境幽雅，拟建床位1200张，地上22层，地下1层，其中包括门急诊、医技、住院、后勤行政科研教学和健康产业用房等。

本方案充分挖掘地域特质，突出建筑创作的独特性与差异性；强调建筑体块的对比与穿插，平面和立面均体现雕塑感，从而产生独特的富有节奏感的建筑个性；运用体现水墨韵味的色调，即丰富的灰色系作为建筑主要色彩；引入水体，使其与建筑互相穿插、融合，形成特色鲜明、空间形态丰富的现代建筑形象。方案从象征性和地域性出发，打造现代化医疗建筑。整体体型象征着跳动的心电图，体现了医院肩负延续生命的神圣职责。

方案以南苑路为界把用地分为南、北两大功能区，南区为医疗区，北区为行政科研后勤保障区。医疗区设医疗综合楼，包括门诊、急诊、医技、病房等功能区。该地块呈矩形，东西面宽、南北向窄，因此设计将串联各功能区的医疗主街横向布置，同时将医疗综合楼尽量靠场地西侧布置，以留出南侧主入口广场和东侧中心绿地及今后发展的空间。行政科研后勤保障区内设行政科研楼及健康产业区，楼内设置行政、教学科研、厨房、餐厅及值班公寓等用房，并通过地面及地下通道与医疗综合楼紧密联系。

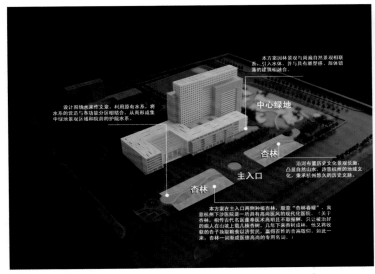

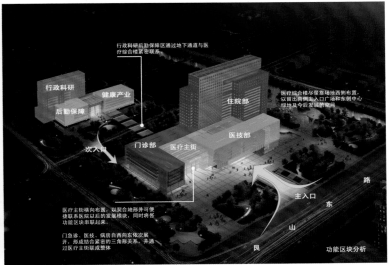

功能区块分析

Liberation Army Jichang Road Branch of No.117 Hospital

Location: Jianggan district, Hangzhou city
Planned and designed by: Zhejiang Modern Architecture Design & Research Institute
Frame: frame-shear wall structure
Land Area: 16 426 m^2
Floor Area: 86 650 m^2
Building Density: 40.8%
Floor Area Ratio: 3.96
Greening Ratio: 30.1%
Parking Space: 300
Number of Beds: 800
Storey: 15

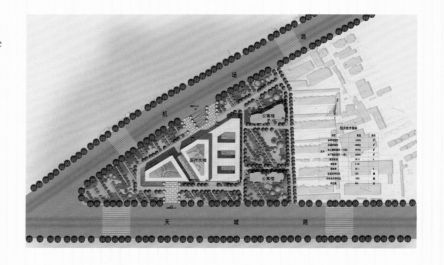

Jichang Road Branch of Chinese PLA 117 Hospital is located in Jianggan District of Hangzhou within the triangle plot in the intersection between Jichang Road and Tiancheng Road. The land is not well-off, and the shape is irregular. The design takes full advantage of the existing layout using centralized approach to distribute the constructions all over the base. All the functions of the hospital are collocated within a medical building composed of three underground floors, 15 ground floors and 5-storey podium.

The design combined with the special temperament of military hospital and military personnel makes full use of the geographical advantage of the road intersection to create a magnificent and masculine image. The main building is tall and straight interlacing with the lower podium. Through the cutting treatment to the body of the hospital building and the actual vs. virtual contrast in the material, the elevation style is unified and magnificent, serious and solemn, reflecting the military's masculine and tough charm. The main building uses triangular layout to organically fit the status quo of the land; the central body cuts into the mass so that the actual vs. virtual contrast is more intense to reinforce the sense of sculpture. The main building is placed in the west of the land, through central traffic core dividing the plane into two nursing units. Within the nursing units wards are arranged to the southeast. The podium is located in the east of the land, making a clever use of the base's long side to arrange all the out-patient terminals in order to meet the lighting. At the same time, the podium is used as a transitional space to draw away the main building from the apartment building at the right side. The underground floors of the medical building link with the subway transfer platform so that patients can enter the hospital directly from the subway station. The design sets up greening roof on the main building and the podium and combines the urban greening landscape system with the hospital greening, thereby sharing the urban landscape as well as providing a more comfortable natural environment for the patients.

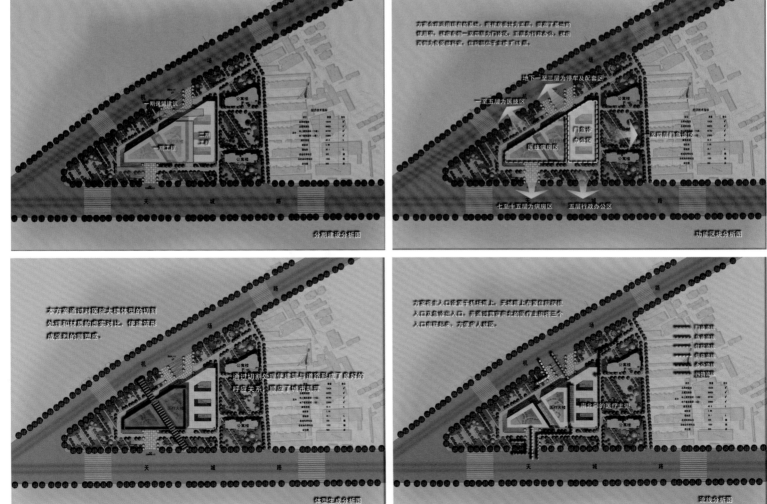

解放军一一七医院机场路分院

项目地址：杭州市江干区
规划/建筑设计：浙江省现代建筑设计研究院有限公司
建筑结构：框架剪力墙
总用地面积：16 426 m²
总建筑面积：86 650 m²
建筑密度：40.8%
容积率：3.96
绿地率：30.1%
停车位：300
总床位数：800
建筑层数：15

一一七医院机场路分院位于杭州市江干区，机场路与天城路交叉口的三角形地块内。用地并不宽裕，形状又不规则。方案充分利用现有用地采用集中式的布局方式，将建筑铺满整个基地。医院所有功能集中布置于一栋医疗大楼内。大楼地下3层，地上15层，裙房5层。

方案结合部队医院及军人的特殊气质，充分利用道路交叉口的地利优势，创造出一个大气、阳刚的外观形象。主楼高耸挺拔，与裙房体型高低错落。通过对医院大楼体型的切割处理和材质的虚实对比，立面风格统一大气，严谨庄重，体现出军人坚毅硬朗的阳刚之美。主楼采用三角形布置形式，有机地契合用地现状。主楼中部切入体块，使主楼的虚实对比更加强烈，强化了建筑的雕塑感。主楼布置于用地西侧，通过中部交通核心将平面分为左右两个护理单元。护理单元内病房均布置于东南向。裙房则位于用地东侧。巧妙利用基地的长边布置各门诊尽端。满足门诊对日照采光的需求。同时，利用裙房作为过渡空间，拉开主楼与右侧公寓楼的间距。医疗大楼地下层与地铁换乘平台联系起来，病人可直接从地铁站进入医院。方案在主楼及裙房屋顶设置绿化种植屋面，并将城市绿化景观体系与院区绿化相结合，共享了城市景观的同时也为病人提供更加舒适自然的住院环境。

Linchuan Cultural Garden

Location: New City Area, Fuzhou City
Planned and Designed by: Beijing Sino-sun Engineering and Consulting Limited

Land Area: 134 ha
Floor Area: 23 740m²
Floor Area Ratio: 0.177
Building Density: 6.7%
Greening Ratio: 50.2%
Parking Space: 330

Object of Fuzhou Linchuan cultural garden planning is to build it the base and center of celebrity culture, history and culture, folk culture, religion culture, essence of drama and cultural displaying, collection and research. It combines structures of Chinese traditional spatial pattern, contemporary public theater, museums, libraries and other buildings together while emphasizing the spirit and atmosphere. On the other hand, since Fuzhou Linchuan Cultural Park is an important node of the whole new area axis, standing against the New District Executive Plaza lengthways, there is an open space settled, clearly highlighting the spatial axis. Overall style of cultural square is a large-scale modern and antique garden square with abundant cultural atmosphere.

It takes Chinese traditional space layout for reference during the design to create space and render atmosphere. When disposing the axis, it considers that land shape in the south is radial, so the combo of museum and library, Xianzu Tang Grand Theater shall be processed into different shapes of structure interfaces to connect two different forms of space, which also makes it become the landmark building in the area.

Designing approach is to closely blend the landscape gardens with strong Chinese cultural characteristics, the traditional courtyard enclosed space and modern buildings together. Sense of space brought by the enclosed landscape structures is applied into the integration and exchanges of architecture and culture. While the half enclosed public space is forged by the large-scale echoing of new buildings. It is extended with two main entrances, respectively forming the longitudinal landscape space axis and latitudinal culture of axis.

The whole square is consist of four scenery walls reflecting Linchuan culture, central axis plaza and the chief sculpture, which contains Linchuan history record wall, bel-esprit biography wall, folk backdrop wall and poetry painting wall. Four large plazas are enclosed by surrounding structures combined with them, there is the Linchuan bel-esprit biography garden reflecting the celebrity, Children-elders garden for leisure, folk culture open-air performances square in the south and Chinese traditional culture lyceum.

Landscape planning is not only stressing the strict spatial order, it also creates a warm, comfortable campus landscape combined with China's infiltration of gardens and green spaces in southeast China at the same time. We have taken representative of spatial form for Chinese traditional landscape features for reference in the design, disposing the transition space around libraries, museums and theaters into a varied green landscape. The garden architectural idea may organically combine the complex with strong sense of the times together; It not only absorbs the essence of Chinese garden culture well, but also deliberately imitates traditional architectural pattern and style, creating a space full of cultural heritage and modern features.

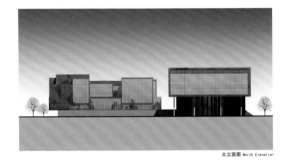

北立面图 North Elevation

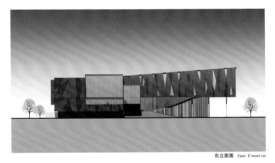

东立面图 East Elevation

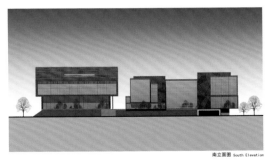

南立面图 South Elevation

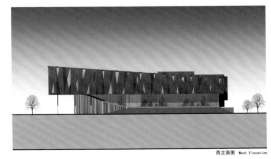

西立面图 West Elevation

临川文化园

项目地址：抚州市新城区
规划建筑设计：北京东方华太工程咨询有限公司

规划用地面积：134ha
总建筑面积：23 740m²
容积率：0.177
建筑密度：6.7%
绿化率：50.2%
停车位：330

抚州临川文化园规划设计的目标是把它打造成抚州名人文化、历史文化、民俗文化、宗教文化、戏曲文化精华的集中展示、收藏、研究的基地和中心，在强调其精神和氛围的同时把中国传统空间格局和当代公共剧场、博物馆图书馆等建筑相结合；另一方面，由于抚州临川文化园是整个抚州新区中轴的一个重要节点，与新区行政广场南北相呼应，因此设计开放型空间布局，明确突出空间轴线。同时文化广场的整体风格为具有浓郁临川文化气息的现代与仿古的大型园林广场。

设计时用中国传统的空间布局来营造空间、渲染气氛。在重点处理轴线的同时，考虑到南侧用地呈放射状，把博物馆与图书馆联合体及汤显祖大剧院处理成东西不同形状的建筑界面，以联系两个不同形式空间，同时也使之成为该区地标性建筑。

规划设计的手法是把具有较强的中国人文特色和园林景观，传统合院围合的空间和现代建筑紧密地揉和在一起，在处理建筑与广场文化园的融合、交流时，利用景观构成墙围合空间感，通过新建筑在更大尺度上的呼应，营造半围合式的公共开放空间，并把它与两个主入口延长，分别形成南北向的主体景观空间轴线和东西向的文化轴线。

整个广场由四道反映临川文化的景墙和中心主轴广场及主题雕塑共同架构而成，分别是临川志史墙、才子传记墙、曲艺背景墙和书画诗歌墙，同时结合建筑围合出四大主题广场，有反映临川名人的才子园、供群众休闲的童叟园、南侧曲艺文化露天演出广场、东侧的国学园。

景观规划在强调严谨空间秩序的同时，结合中国江南园林绿化空间的渗透，营造温馨、舒适的景观校区。设计时参考了具有中国代表传统园林景观特点的空间形式，把围绕博物馆图书馆和剧场的过渡空间处理成变化丰富的绿化景观，造园理念把具有强烈时代感的建筑群有机的融合在一起，既很好的吸取了中国园林文化的精髓，又不刻意模仿传统建筑格局和造型，从而形成了一个具有文化底蕴又具有当代特色的空间。

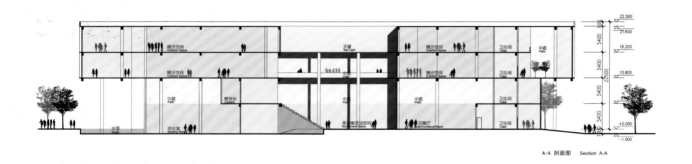

A-A 剖面图 Section A-A

B-B 剖面图 Section B-B

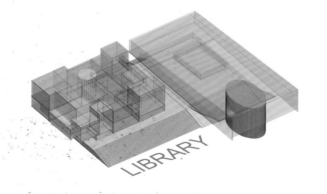

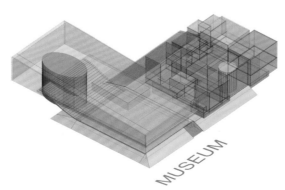

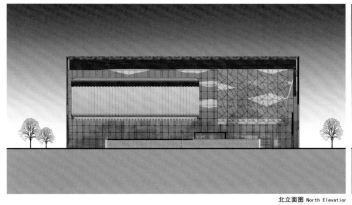
北立面图 North Elevation

南立面图 South Elevation

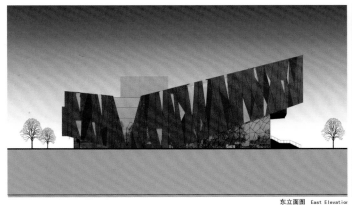
东立面图 East Elevation

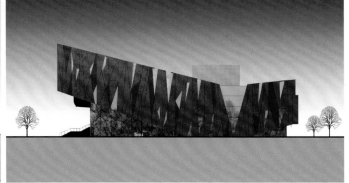
西立面图 West Elevation

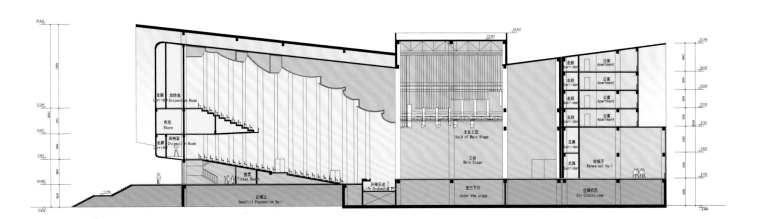

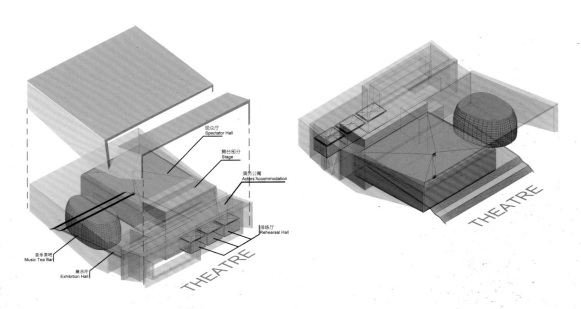

Guangzhou Children's Palace of Luogang District

Location: Education Region of Luogang Central District, Guangzhou city
Planned/Designed by: Guangzhou Kecheng Construction Design Co., Ltd

Land Area: 19 902m²
Floor Area: 21 080m²
Building Density: 24.8%
Floor Area Ratio: 1.06
Greening Ratio: 38.2%
Parking Space: 116

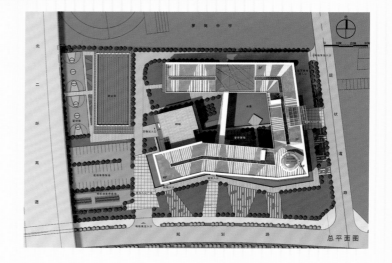

西南夜景鸟瞰图

Luogang Children's Palace is proposed to locate in the education region of central district and will be capable to accommodate 4000 people for trainings at the same time after it is established. The base is flat with convenient traffic around, but Northern Second Ring Road produces many noises in the west.

"You will never know the result of next turn"------this is the reason for the perpetual power of magic wand since its birth. The teenage are just the group with unknown secrets and huge potential while Luogang Children's Palace is just like an attractive magic wand and the place for simulating the potential of the teenage and cultivating their personalities and showing their energy. Designers adopt a kind of rising technique to shape the construction and organically unfold it inside and combine it with cinema body to form a strong comparison and abundant gray space blending outside and inside. The facade of building is composed by a large perforated metal plate and a vertical shutter which wreathe each other to embody a kind of hi-tech sense and also form a uniform wholeness with the shape of construction. Meanwhile, it can also shield the parts of external air-conditioners from affecting the beauty of construction, contribute to energy-saving purpose and reduce the influence on construction from environment. The facade facing internal yard is built by open glass to integrate environment into the construction. Designers closely combine the appearance of building with energy-saving measures and space of sight to make a remarkable landmark in the district. The construction adopts the concept of "human intercourse space inserting concept". Intercourse platform and roof garden are designed to be the center of inside, leisure and sight to let construction space and environmental space get into each other and become a wholeness. Corridor and intercourse platform are adopted in each storey to create a wonderful communication space to achieve the purpose of convenience and communicate with the environment.

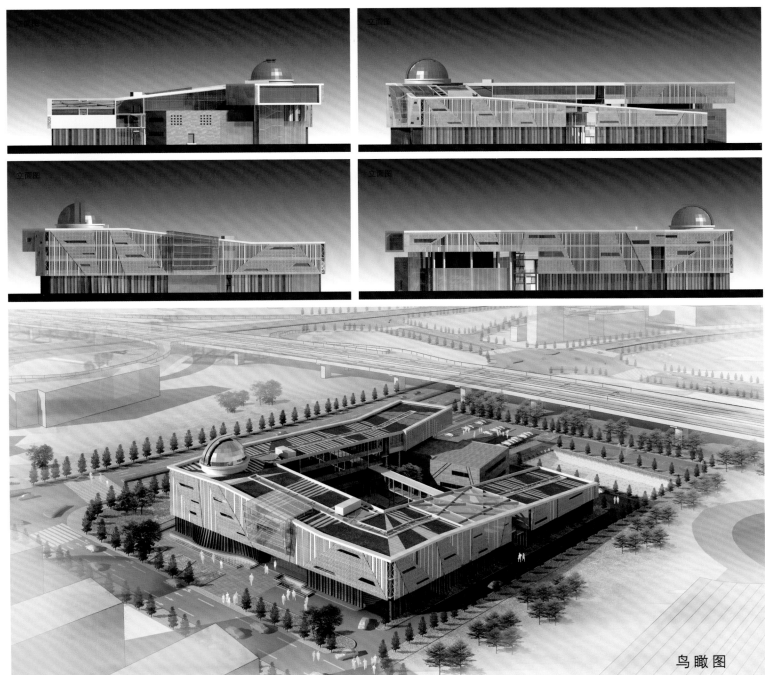

鸟瞰图

西南低点透视图

广州市萝岗区少年宫

项目地址：广州市萝岗区中心区教育片区
规划/建筑设计：广州市科城建筑设计有限公司

占地面积：19 902㎡
建筑面积：21 080㎡
建筑密度：24.8%
容 积 率：1.06
绿 化 率：38.2%
停车位：116

萝岗区少年宫拟建于萝岗中心区教育片区，项目建成后可同时容纳4000人培训。基地地势平坦，四周道路交通方便，西面北二环高速路噪声影响较大。

"你永远不知道下一次转动会有怎样的结果。"这是魔棍从诞生之初到现在一直长盛不衰的原因。青少年所具有的正是这样一种充满无尽未知数，暗含巨大潜力的特质，而萝岗区少年宫正如同一根充满无穷魅力的魔棍，成为激发青少年内在潜力，培养其个性并展现其活力的一处场所。建筑总体造型采用由低到高逐渐升起的手法，在用地内有机地舒展开，并与剧场体块穿插组合，产生强烈的对比，形成丰富的室内外交融灰空间。建筑外立面为大面积穿孔金属板与竖向百叶相互咬合环绕，体现出一种高科技感，但又同时与建筑形体形成统一的整体。同时也可以遮挡外部空调机等有碍建筑美观的零散构件，起到节能的作用，并减少外部环境对建筑的影响。面向内部庭院的立面则采取较为开敞的玻璃面，使环境融入建筑内部之中。建筑造型与节能措施、景观空间的需要紧密结合，成为该地区中令人过目难忘的地标。建筑设计采用"人性交往空间穿插概念"，其中设置了交往平台和屋顶花园作为建筑内部与休闲及景观中心，使建筑空间与环境空间相互交融，成为一个整体。各层运用连廊与交流平台，创造丰富的人性交往空间，到达便利，充分与环境景观对话。

一层平面

二层平面

三层平面

四层平面

内庭院透视图二

内庭院透视图一

Huidong Old Cadre Activity Center

Location: Urban Area of Huidong Town, Huizhou City, Guangdong Province
Architecture Design: Guangzhou Keycity Architecture Design Co., Ltd

Land Area: 13 969.3 m²
Floor Area: 9 179.8 m²
Building Density: 23.5%
Floor Area Ratio: 0.66
Greening Ratio: 46.3%
Parking Space: 207

Architectural space may have great difference in the quality and charm which is relying on the availability of artistic rendering. Traditional Chinese architecture usually attaches more importance to the amalgamation balance between "reality" and "fantasy", which is to say it is more careful about the balance connection among the constructions in site and other elements, as well as their amalgamation location in modal. The so-called deep courtyard is describing that Chinese courtyard is like to open never-ending small cases, one after another, entering one yard then next, just like walking in a roll of long banner painting. Buildings around the courtyard take the inside yards as link space, which is of publicity

nature, effectively organizing stream of people. Therefore it becomes a collector-distributor space of convenience, efficiency and broad vision. The most important advantage is the perfect cortile with ecological significance, which is always bathing in the sunshine and cuddling to the virescence. Inside yard could bring energy to construction - the sunshine and air. That is what the design is seeking for: the elder based; tender and pleasant size; multiple static- spaced yard; sunshine and tree shade.

Site characteristics - elevation difference, connection with plaza and streetscape built with the flow. It not only has created a living environment full of fun and charm, but also effectively saved investment.

Traditional mountain architecture has created a great abundant space handling technique in the settlement of vertical relationship between constructions and topography. Buildings go with the landform up and down, or constructing platforms, increasing plinth, or by means of ladder-shaped foundation, split-level houses and height going with landform, or applying overhang, rock attaching, overhead and so on. A combination use of these techniques can save the amount of earth and stone, maximizing the constructing space, which dynamically combined buildings with landform together.

Sunshine garage is introducing the natural ventilation and lighting in, which is not only beautiful, comfortable and eco-environmental protective, but also saving the cost.

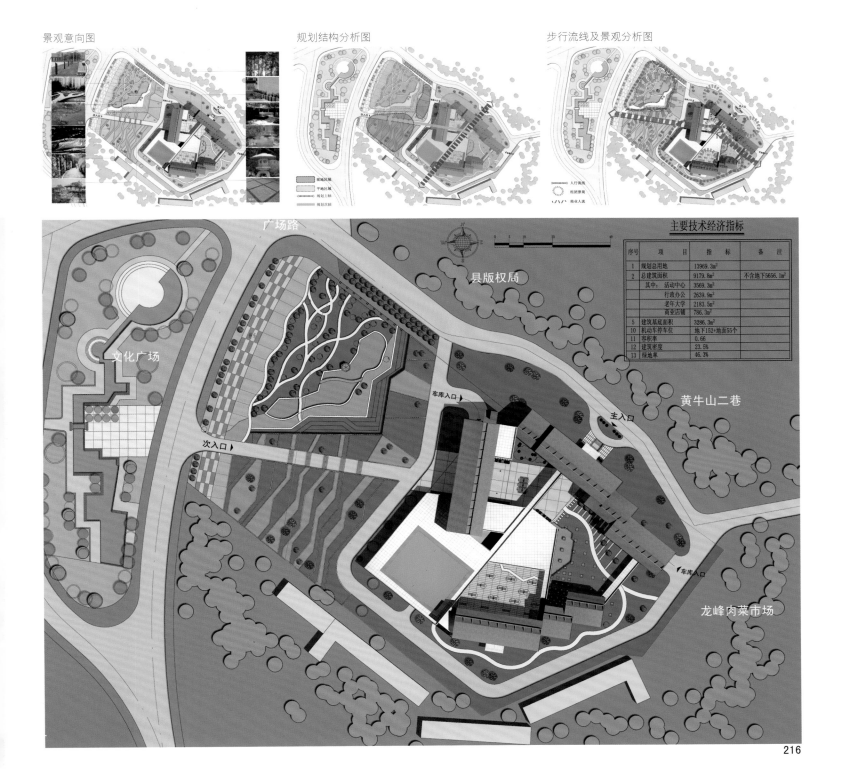

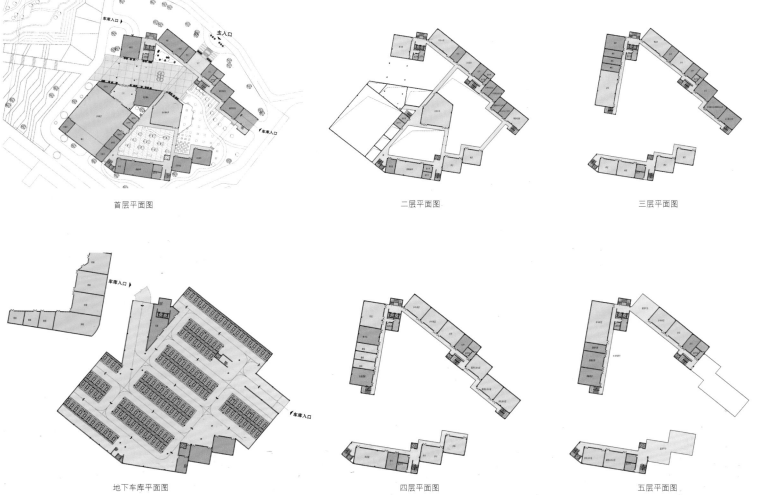

| 首层平面图 | 二层平面图 | 三层平面图 |

| 地下车库平面图 | 四层平面图 | 五层平面图 |

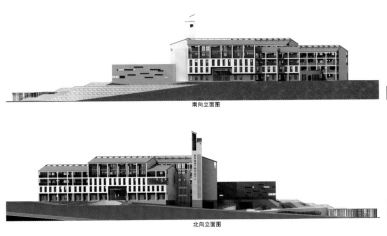

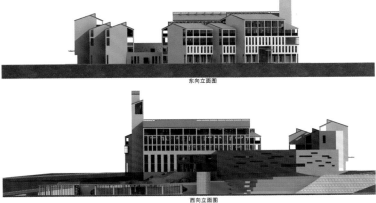

惠东县老干部活动中心

项目地址：广东省惠州市惠东县市区
建筑设计：广州市科城建筑设计有限公司

占地面积：13 969.3㎡
建筑面积：9179.8㎡
建筑密度：23.5%
容 积 率：0.66
绿 化 率：46.3%
停车位：207

有无意境之渲染，建筑空间的品质境界大不相同。中国传统建筑"虚"、"实"之间的平衡融合，也就是重视场地中建筑物与其他要素之间在位置上的平衡关系和形态上的融合关系。庭院深深表达的是中国庭院好像开启不完的匣子，开完一个又一个，进入一个院又进一个院，好像走在一卷横幅画轴里。周围的建筑以内庭院作为联系空间，呈现公共性，有效组织人流，成为高效便捷、视野开阔的集散空间，至关重要的好事具有生态意义的中庭，始终沐浴着阳光，拥抱着绿化。内庭院带给建筑能量——阳光、空气，这便是设计所追求的：以老人为本；亲切、怡人的尺度；多静态空间——院落；阳光与树荫。

场地的特征——高差、与广场的对话、街景顺势而筑——发挥用地条件优越，顺势而筑，既能营造富有情趣意境的居住环境，又能有效地节约投资。

传统的山地建筑在解决建筑与地形的竖向关系方面，创造了极丰富的空间处理手法，建筑随地形的高低起伏，或筑台、提高勒脚；或掉层、错层、跌落；或悬挑、附岩、架空等等。综合运用这些手法，可以节约土石方工程量，争取建筑空间，使建筑与地形有机地结合。

阳光车库引入自然通风和采光，既舒适美观，又生态环保，也可以节约工程造价。

Science & Culture Center of Yunnan Chihong Xinzhe Incorporated Company

Location: Qujing City, Yunan Province
Designed by: German S.I.C-Engineering Consulting Limited Liability Company
Designer: Tim Bonnke

Land Area: 120 300 m²
Floor Area: 19 800 m²

"Science & Technology Cultural Center" is the core agency of Chihong Company, integrating commercial activities and service function together such as office, scientific research, staff training e.t.c. There will be offices, R & D center and training center for employees. Main office building is the focus of this program in modern, unique and chic style. The design inspiration comes from the mine and the seam. Vertical core transportation tube (elevator access) inside the construction is on behalf of the mine, while the horizontal-oriented unevenly spreading floors are standing for seams. The subject is very suitable for the company's business direction of mineral resources development.

Cantilevered parts of the building shall emerge different lighting effect with the changes of sun angle and intensity in a day. The overall design reflects very high artistic quality and forward-looking nature, which also reflects the company's business philosophy "high-tech to drive the production, innovation to achieve transmutation". Zinc is used as the main decorative material for outer surface of the building, while "zinc" is also the company's main product, which is undoubtedly the perfect interpretation of the company's product. The unique architectural form can be used as the best advertisement to promote the company image.

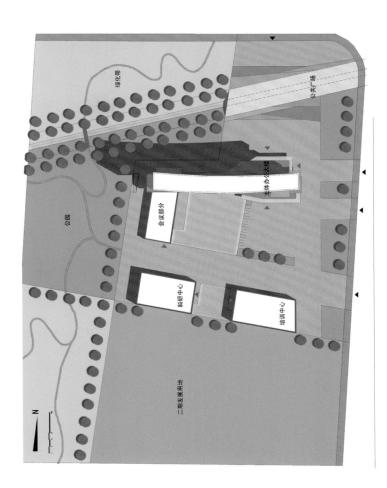

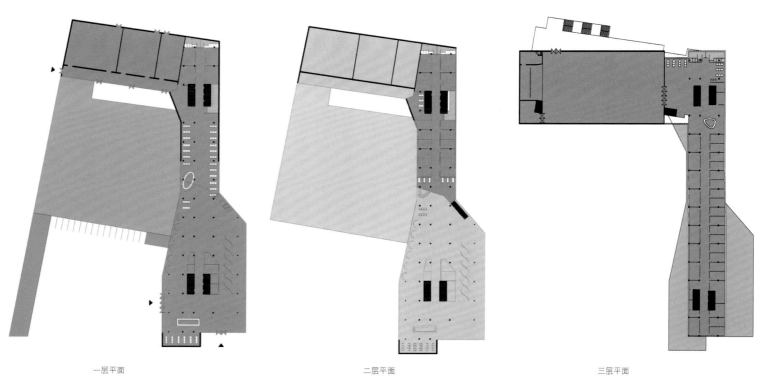

一层平面　　　　　　　　　　　　二层平面　　　　　　　　　　　　三层平面

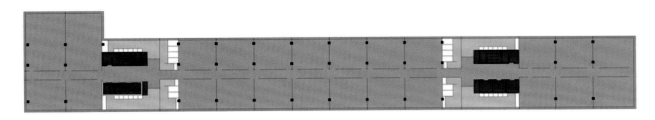

十三/十四层平面

十一/十二层平面

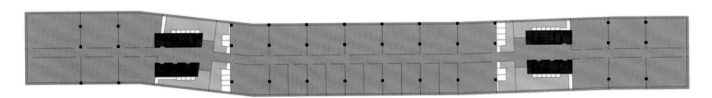

九/十层平面

云南驰宏锌锗股份有限公司"科技文化中心"

项目地点：云南省曲靖市
设计：德国S.I.C.-工程咨询责任有限公司
设计师：蒂姆·邦克 (Tim Bonnke)

基地面积：120 300 m²
总建筑面积：19 800 m²

"科技文化中心"是驰宏公司集办公、科研、职工培训等商务活动和服务功能为一体的核心机构。设置了办公中心、研发中心以及职工培训中心。办公主楼是本方案的重点，其风格现代、造型独特而别致。设计灵感来源于矿井及矿层。位于建筑内部的竖向交通核心筒(电梯通道)代表着矿井，而水平方向参差不齐的楼层则代表着矿层。这个主题与该公司矿产资源开发的经营方向非常的切合。

随着一天之中太阳照射的角度和强度随时间不断变化，建筑的悬挑部分能产生不同的光影效果。整体设计体现了极高的艺术性和前瞻性，同时也体现出该公司"高科技带动生产，创新实现腾飞"的经营理念。采用锌材料作为建筑外表面的主要装饰材料，而"锌"也是该公司的主要产品，这无疑是对该公司产品的完美诠释。建筑形态的独一无二性，可作为企业形象推广的最佳广告。

Hefei Federation of Trade Unions Staff Activity Center

Hefei Federation of Trade Unions Staff Activity Center
Project Address: Hefei
Designed by: UDG (United Design Group)

Land area: 14 000 ㎡
Floor area: 37 000 ㎡

The Complex Building Project for Staff Activity Center of Federation of Trade Unions of Anhui Province is located in Hefei Political & Cultural Center, the south-east corner of the junction between Qianshan Road and Jiahe Road. With 21 floors on the ground and one floor underground and framework shear wall structure, the design combines and splice the simple and magnificent masses to form an extraordinary program.

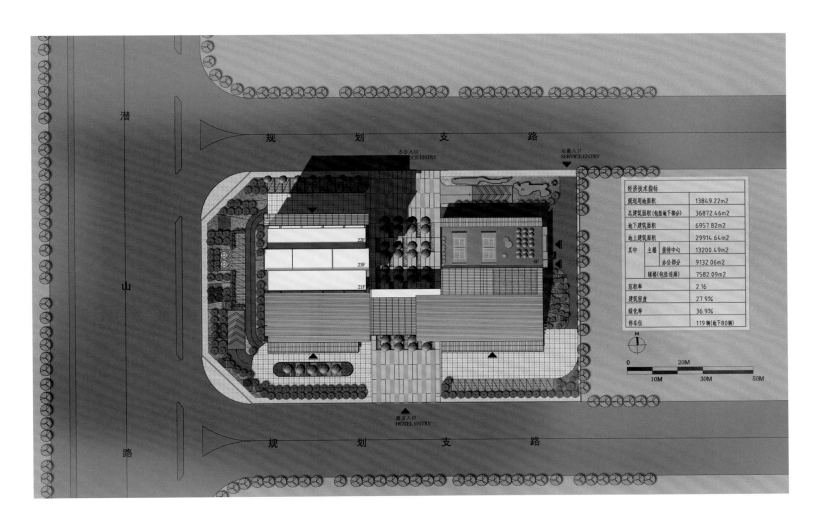

鸟瞰效果图

合肥总工会职工活动中心

项目地址：合肥
方案设计：UDG联创国际

用地面积：14 000m²
总建筑面积：37 000m²

安徽省总工会职工活动中心综合楼工程位于合肥市政务文化中心潜山路和嘉和路交叉口东南角，地上二十一层，地下一层，框架剪力墙结构，设计以简约、大气的体块拼接组合，形成不同凡响的方案。

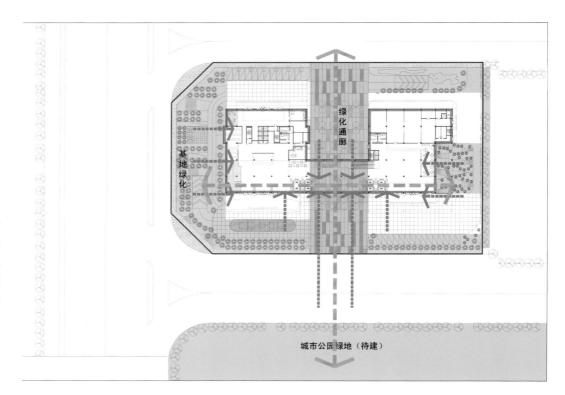

日景效果图

大堂效果图

日景效果图

New Teaching Building of College of Physics, University Rostock

Location: Rostock, Germany
Designed by: German S.I.C-Engineering Consulting Limited Liabilities Company
Designer: Jan Gutermuth

Land Area: 5 400m²
Floor Area: 6 687 m²

The contest is the new building project of physics department, University Rostock. The new building will rearrange the buildings of original physics department which are disseminated in the city. The building will be located in the development land for university, south of Einstein, Albert Street, in the south of Rostock city.

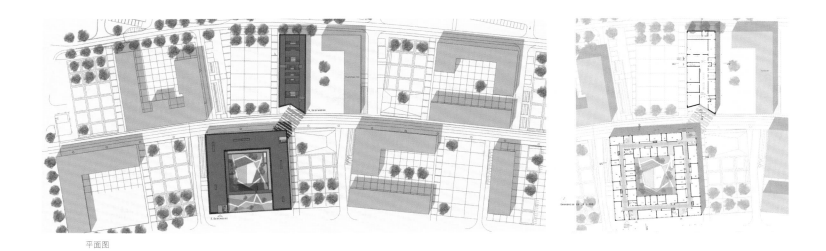

平面图

立面图

立面图

立面图

剖面图

罗斯托克大学物理学院新教学楼

项目地点：德国罗斯托克
设计：德国S.I.C.-工程咨询责任有限公司
设计师：杨·古特木特(Jan Gutermuth)

总用地面积：约5 400m²
总建筑面积：6 687 m²

　　该竞赛是罗斯托克大学物理系大楼新建项目，新建的系馆是将散布在该城市的原物理系各建筑重新整合。新建大楼将坐落于罗斯托克南城的阿尔伯特·爱因斯坦大街南侧的大学发展用地上。

Foshan Complex of Public Culture

Location: Foshan city, Guangdong province
Designer: Qiu Huikang, Chou Chaohui, Enrico, Sun Xiaofei, Xiaoming
Scale: 40 000m²

Foshan Cultural Port is a real urban complex which can realize the integration of functions, matching facilities and traffic combining man, vehicle, watercraft and rail transportation into this great structure to support a 3-D and composite development. Designers hopes that when Foshan Cultural Port, this grand "urban complex" is totally complete, it will integrate cultural entertainment, social intercourse, public service, business, leisure entertainment into one to form an coverall Foshan.

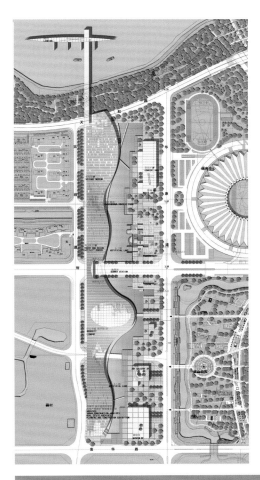

At the bottom of main functional bodies sets a street-style space going through south to north. It is a mall, a cross-cultural mall which is composed by public service, business, bookstore and LOFT. Like a functional chain, it organically relates each separate functional body together not only guaranteeing the separate independence and facilitating the use and management, but also providing a reasonable schedule for the staging development of complex.

The Cultural Port will be completed to have the environment of a park and the open window of the city. The project will realize the integration of construction and environment. When it is complete, it will organically consolidate Foshan Eco Park, Foshan Sports Park and Foshan Binjiang Park to form a Foshan central park with complete urban functions, diverse forms and geographical characteristics.

Designers expect the future Foshan cultural complex will be a great structure, dock, harbor, air port and the communication platform of urban culture---- a cultural port. The Dongping River will become the eastern Seine River in the future and the cultural complex will be a cultural dock on the river.

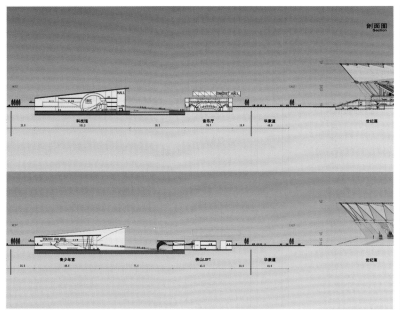
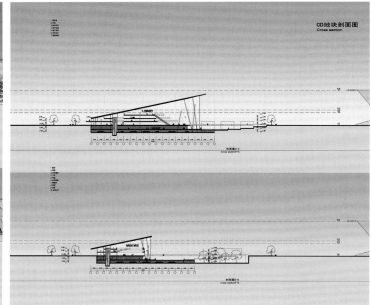

佛山公共文化综合体

项目地点：广东佛山
建筑设计：深圳立方建筑设计顾问有限公司
设计人员：邱慧康、仇朝辉、Enrico、孙晓非、肖明
项目规模：40 000m²

佛山文化港是一个真正的城市综合体，能够实现功能、配套和交通的综合，将人、车、舟和轨道交通在这个巨构里实现一体化设计，立体复合式发展。设计期待佛山文化港这座规模宏大的"都市综合体"全部落成后，将集文化娱乐、社交、公共服务商业和休闲娱乐等城市功能于一身，成为一个包罗万象的"佛山城中城"。

在主要功能体的底部设置的一条贯穿南北的街道式空间的是一个MALL，是以公共服务商业、书城、LOFT等功能组成的泛文化的MALL。它就如同一条功能链，将各个相对独立职能管理的功能体有机的连接在一起，又保证了各自的独立性，便于使用和管理，而且为综合体的分期开发提供了合理进度。

文化港的建成环境是公园的环境，是城市开放的窗口，实现了环境与建筑的一体化。它的实现能够将现有的佛山生态公园、佛山体育公园和佛山滨江公园有机地整合成一个具备完全城市功能，形态丰富，具有地域特色的佛山中央公园。

设计师期待未来的佛山市文化综合体是巨构，是一个码头、一座港湾、一个空港、一个城市文化的交流平台—佛山文化港。未来的东平河将成为东方塞纳河，而文化综合体就是河上的一个文化码头。

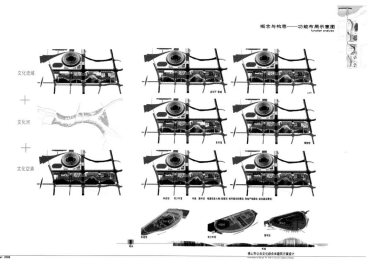
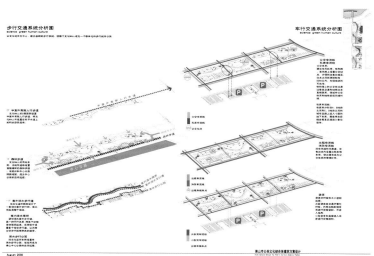

Teacher's studio of Sichuan Fine Arts Institute

Planned and Designed by: Dblant Design International
Land Area: 27 947.4m²
Floor Area: 54 320.3m²
Floor Area Ratio: 1.94
Building Density: 45.9%
Greening Area: 8407.4m²
Greening Ratio: 30.1%
Parking Space: 60

The foundation of teacher's studio of Sichuan Fine Arts Institute is located in the east side of its Huxi Campus with municipal roads in the east and uplands with gentle slope in the west. The scope of the foundation forms a rectangular plot with a length of 298.4m in the south-north direction and that of 93.66m in the east-west direction. It is high in the west and low in the east with a relatively large inclination in the east-west direction; it is high in the south and low in the north with a relatively slight inclination in the south-north direction. There is neither buildings reserved within the plot nor main municipal pipelines passing through. This project primarily functions as the fine arts studio for the teachers of the institute.

Design feature: combined with the current situation of the foundation, the centralized-style layout is adopted and added with several lighting atriums in different sizes in the center. The symmetric capacity is employed and the interspersing and staggering shadow effect among blocks is highlighted. The building is placed from south to north in a long shape, in the middle part of which a channel divides the building into the eastern section and the western section which are connected through corridors. An open space for sharing is set up in the middle of the building, forming several independent capacities which are both connective and dependent. Combined with the original topography, we use as less earth volume

透视图

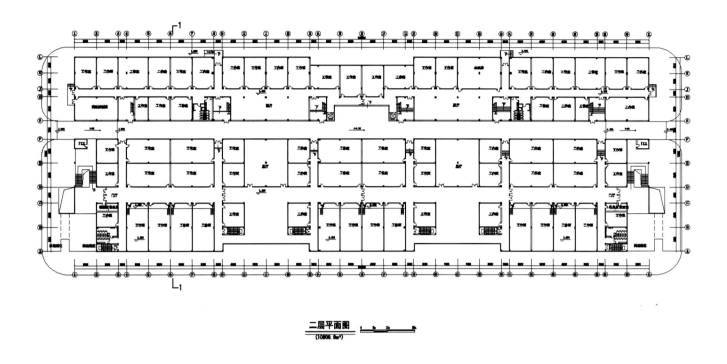

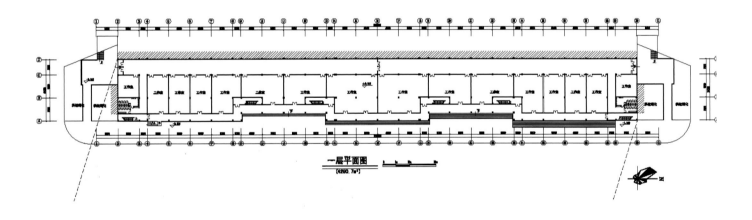

as we can during the process of design and the building is designed to be a platform-type one. The landscape steps and the platform are arranged at the south and north side of the building and the main entrance is set above the platform. On both sides of the steps is the ramp for vehicles, the inclination of which meets the standard requirement. The trend of the rainwater within the plot conforms to the aspect of the slope it drains from the inside out to the city roads. The architectural style of the studio of the institute is in accordance with the artistic and cultural atmosphere of it. The connecting and interspersing of the main materials are matched up the changing of the capacity of the building itself to create an experience of outward appearance with plentiful layers. Walls made of different materials are matched with the window types of different layouts, which further strengthens sense of wholeness and sense of rhythm of the facades.

四川美术学院教师工作室

规划建筑设计：都林国际设计
总用地面积：27 947.4㎡
总建筑面积：54 320.3㎡
容积率：1.94
建筑密度：45.9%
绿地面积：8407.4㎡
绿地率 ：30.1%
机动车位：60

四川美术学院教师工作室基地位于四川美术学院虎溪校区东侧，东临市政道路，西临缓坡山地。基地范围是长方形块地。基地南北长298.4m，东西长93.66m。基地西高东低，南高北低，东西向坡度较大，南北向坡度小，场地内无保留建筑，也没有主要市政管线穿过。该项目的主要功能是美院教师的美术工作室。

设计特点：结合基地现状，采用集中式布局，中间加入大小不一的多个采光中庭。造型采用左右对称的体量，强调体块间的穿插与交错的光影效果。依据基地走向，建筑呈长边南北向布局，中部有通道将建筑划分为东西两部分，两部分间通过连廊相连。建筑中间还设有开敞的共享空间，形成几个既相连又独立的建筑体。设计中我们结合原始地貌，尽量减少土方量，设计为台地式建筑。建筑的南北端设置景观台阶和平台，主入口设在平台之上。台阶两侧是车行坡道，坡度满足规范要求。场地雨水的走向则顺应场地原来的坡向，由内向外排向城市道路。美院工作室的建筑形式和美院的艺术、文化氛围相适应，以红砖为主要材料的交接与穿插，配合建筑体量本身的变化，营造出有丰富层次的外观体验。不同材料的墙面配合不同构图的开窗形式，更加强化了立面的整体感和韵律感。

透视图

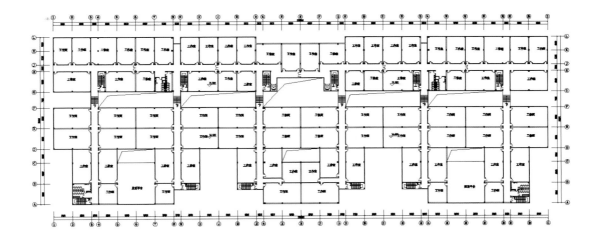

四层平面图

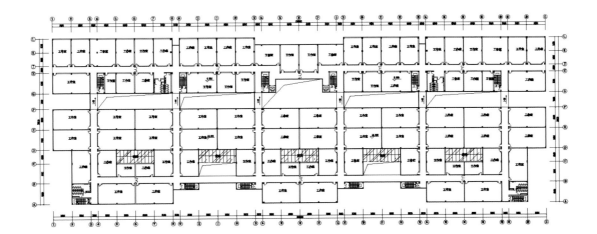

三层平面图

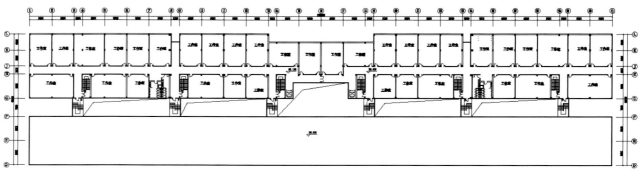

六层平面

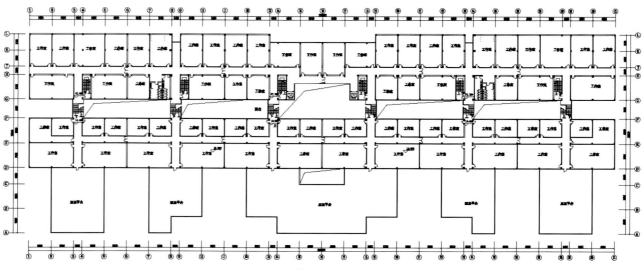

五层平面

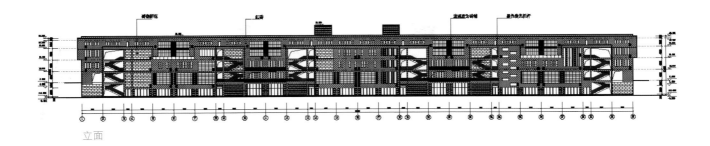

立面

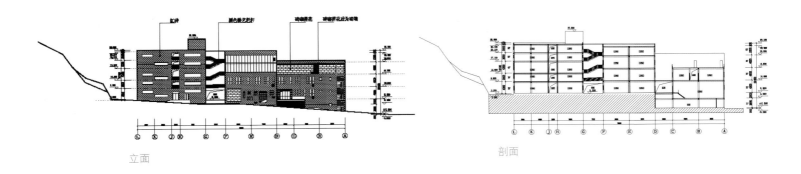

立面　　　　　　　　　　　　　剖面

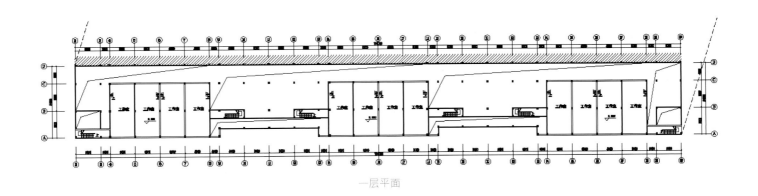

一层平面

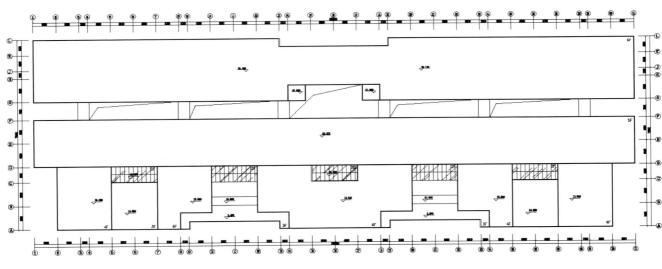

屋顶平面

URBAN PLANNING

城市规划

Conceptual Planning of Binghu New Town, Dongtaihu Lake, Wuzhong, Suzhou City

Conceptual Planning of Binghu New Town, Dongtaihu Lake, Wuzhong, Suzhou City
Planned and Designed by United Design Group (UDG)

The present Wuzhong not only possesses the business prosperity with Wuzhong International Business City and Suyuan Restaurant as representative but also possesses many uprising new towns which are featured by the characteristics of Jiangnan Water Town, rich cultural context and modernized atmosphere. Magnificence, elegance, fashion and energy have become the new pronouns of Wuzhong new towns.

Binghu New Town is located in the southwest of central city and the Shore of Dongtaihu Lake, with planned area of 10.03 square kilometers and the population of 120 thousand. Binghu New Town will attach importance to introducing star-level hotel, cultural entertainment, medical service, sports and leisure, business and finance, real estate, urban park and characteristic town. Development target is constructing southern Suzhou into a satellite city with the integration of commerce & finance, culture & entertainment, residence & business starting, zoology and tourism, leisure and vocation into one. With coastline of Dongtaihu Lake as a axis, which government is trying great effort to improve, the design will make full use of the landscape of Dongtaihu Lake coastline and build up a water-friendly and harmonious ecological district.

AERIAL VIEW OF URBAN DESIGN

苏州吴中东太湖滨湖新城概念规划

规划设计：UDG联创国际

现今的吴中不仅有以吴中国际商城、苏苑饭店为代表的商业繁荣，还有一批批正在崛起的既有江南水乡特色、文化底蕴深厚，又富有现代化气息的新型小城镇。气派、典雅、时尚、活力正成为吴中新城的代名词。

东太湖滨湖新城位于中心城区西南部，东太湖之滨，规划面积10.03km²，人口规模12万人。东太湖滨湖新城将重点引进星级酒店、文化娱乐、医疗卫生、体育休闲、商贸金融、房地产业、城市公园、风情小镇等项目。发展目标是建设成为苏州城南一个集商贸金融、文化娱乐、居住创业、生态观光、休闲度假于一体的卫星城市。滨湖新城以政府正大力治理的东太湖岸线为轴，充分发挥东太湖岸线的山水优势，建设起一座亲水型、和谐型生态新城区。

Mongolia Wuhai CBD Layout

Location: Wuhai city, Mongolia Municipality
Designed by: German S.I.C-Engineering and Consulting Limited Liabilities Company
Designer: Wukai, Stephan Klabbers, An Jijun, Chen Jiayu

Land Area: 446 000 m²
Floor Area: 1 135 837 m²
Building Density: 29.10%
Floor Area Ratio: 2.55
Greening Ratio: 35.50%

Overall starting point of CBD is to construct a 3-D ecological business central area. In the plane layout, the design advocates introducing green elements into each building. Besides using the large scale landscape greenbelts to vertically go through each business central area, designers value the comfort of each cluster and building by inserting greenbelt between constructions to make sure landscape is accessible from each corner. In the treatment of facade, the design jumps out of common frame by sinking some clusters and raising some, with overbridges and roof gardens connecting them, and completely embodying landscape and office space.

The construction is composed by enclosed or semi-enclosed blocks to embody the rigorous and generous design technique of the community. Road and greenbelt are designed to break some block structures and create the different conversions between public space, semi-public space and private space. Buildings combine with each other to construct interesting spatial changes.

主轴景观分析图

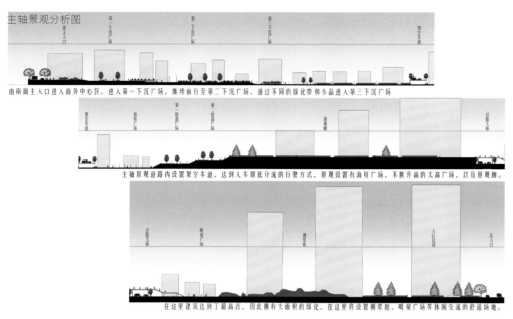

由南面主入口进入商务中心区，进入第一下沉广场，继续前行至第二下沉广场，通过不同的绿化带和小品进入第三下沉广场

主轴景观道路内设置架空车道，达到人车彻底分流的行驶方式，景观设置有海贝广场，不断升高的太高广场，以及景观廊。

在这里建筑达到了最高点，因此拥有大面积的绿化，在这里将设置榴翠庭，喷泉广场等休闲交流的舒适场地。

总平面图

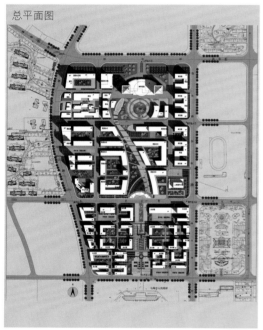

功能分析图

- 市政办公建筑
- 商务办公建筑
- 商住楼
- 餐饮商业
- 酒店公寓
- 文化娱乐

交通分析图

- 城市道路
- 车行主干道
- 人行主干道

景观分析图

- 中心景观轴线
- 商业步行街轴线
- 交流广场
- 点式绿化组团

内蒙古乌海CBD规划

项目地点：内蒙古自治区乌海市
设计：德国S.I.C.-工程咨询责任有限公司
设计师：武凯、史迪文、（Stephan Klabbers）、安继君、陈佳瑜

总占地面积：446 000 m²
总建筑面积：1 135 837 m²
建筑密度：29.10%
容积率：2.55
绿化率：35.50%

CBD总的出发点是打造一个立体式生态商务中心区。在平面布局上设计提倡将绿色引入每座建筑，在利用大面积的景观绿化带竖向贯穿整个商务中心区同时，还注重每个小组团和单体建筑的舒适度，将绿化穿插在建筑当中，保证每个角落都能看到景观。在立面处理上，设计突破一般的单调做法，将地块部分组团下沉，部分抬高，中间用天桥和空中花园连接，将景观和办公空间彻底立体化。

建筑布局呈块状围合或半围合形态构成，体现出整个区域严谨大方的规划设计手法，并利用道路、绿地和小品打破某些方块结构，创造出公共空间，半公共空间以及私密空间的不同转换。建筑体块互相咬合，错落衔接，构建出有趣的空间变化。

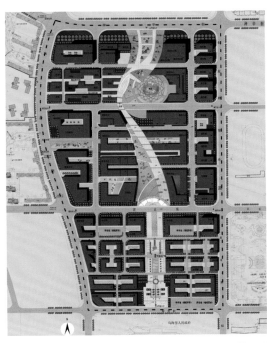

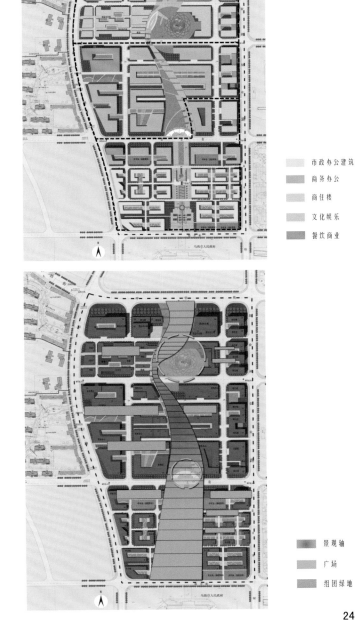

Shenzhen Guanlan Print Base

Designer: Aube Architecture Design (France) Co. Ltd
Land Area: 150 ha
Floor Area: 150 000 m²
Floor Area Ratio: 0.1

This project is aimed to establish a national and even world-class Art Center of Guanlan print. It also has enhanced the cultural image of Baoan District by fully taking advantage of the base's historical environmental resources, which performs as an important business card for Baoan District and Shenzhen.

Print was born in China due to the China's printing technology and printed matter. The protection and promotion of this knowledge have been recognized by fashion trend, which makes it a strong cultural carrier among the wide range of Chinese culture forms. It seems to European collectors that many Chinese young artists are emerging. Shenzhen Baoan's proposal of developing print art base has not only come to be a wish to integrate prints creating, production, trade and teaching together, but also conformed to the international trend.

Location advantages

Advantage 1: golf course. Golf Course not only provides the base with a favorable landscape, but also attracts the attention of many social elites, since many world-class events are held here and live telecasted all over the world.

Advantage 2: locus of the base is under the basic ecological control line's protection which is delineated by Shenzhen Municipal Government.

However, there are still a series of difficult-to-operate matters to achieve this project. Many illegal buildings, factories, new villages and so on emerge on rural land under the impact of chaotic situation in the city, so it is necessary to redefine the land. In fact, the Niuhu community in Guanlan Street, located on the boundary area between Dongguan City and Bao'an District, has always been under the sustained pressure of urban transformation for a long time. The natural structure has completely broken down and gone towards urbanization.

Redefine the base

In such a context, the only way to focus and manage all kinds of wishes and conflicts from the political, economic and social groups is seizing a realistic and ambitious viewpoint. Only the common power will be able to drive the project into success in a series of duplication, mutual adjustment and consummation. Our obligation is to produce a rigorous and rational master plan which could promote a consensus by participants to ensure the feasibility of the project.

深圳观澜版画基地

设计：法国欧博建筑与城市规划设计公司
用地面积：150 ha
建筑面积：150 000m²
容积率：0.1

本项目的目标是要建立一个国家级乃至世界级观澜版画艺术中心，充分利用基地的历史资源和环境资源，提升宝安区的文化形象，使其成为宝安和深圳重要的城市名片。

版画诞生于中国，它的诞生归功于中国的印刷技术和印刷品。这项知识的保护、推广，并被流行趋势所认同，使其成为中国种类繁多的文化形式的强大的文化载体。在欧洲收藏家看来，中国有很多年轻艺术家涌现出来。深圳宝安提出发展版画艺术基地，成为版画创作、生产、交易教学展示于一体的中心的意愿，也顺应了国际。

位置优势

优势之一，高尔夫球场。高尔夫球场不仅为基地提供了良好的景观，而且吸引了很多社会精英的目光，因为在这里经常举行世界级的赛事，并且全球直播。

优势之二，基地处于深圳市政府划定的基本生态控制线保护之下。

然而，实现这个项目需要面对一系列难以操作的问题。

农村的土地在城市杂乱无章的冲击下出现了许多违法的建筑、工厂、新村落等等，需要对这些土地进行重新定义。事实上，观澜街道牛湖社区位于东莞市与宝安区的边界，一直处在长期、持久的城市转型压力之下，自然结构完全崩溃，走向城市化。

重新定义基地

在这样一个背景下，只有拥有一个既切合实际的、又抱负远大的想法和观点，才能够集中和管理来自于政界、经济界人士和社会团体的各种意愿和各种冲突。只有共同的力量才能够有助于项目在一系列的重复和交互调整和完善之中走向成功。设计者的责任就是交出一份严谨的、合理的主体计划，能够让参与者和决策者达成共识，确保计划的可行性。

Qingdao Small Bay Super 5-Star Hotel Design and Coastline Layout

Planned and Designed by: ANS International Architecture Design and Consulting Co., Ltd
Designer: Joe Lau, Yan Deyun, Luzhao

Land Area: 154 000 m²
Floor Area: 128 500 m²
Greening Area: 38 500 m²
Base Area: 26 449 m²
Number of Stories: 69
Building Density: 56.6%
Floor Area Ratio: 3.91
Greening Ratio: 25%

Qingdao Small Bay is located in the east coast of Jiaozhou Bay, south of old urban bay, west of old urban area of Qingdao city, adjacent to Zhonggang, Railway Station and Zhongshan Road and coastline is about 2.8km. The layout reorganizes the structure of coastline following the principle of "one strip, two hearts and two axes", which means the design connects the round arena area in the piazza and main building together by the landscape belt of dock, and waterside urban coast landscape axis and trestle landscape axis link the integral space between human, sea and city to vividly and orderly activate the spatial energy of waterside area.

The design of hotel is going to create the highest structure of Qingdao and the modern urban landmark above 220 meters and make it a new symbol of Qingdao urban civilization. The design of theme hotel and regional coastline layout are blended into one and show the best in each other. Hotel complex provides a series of entertainment and commercial facilities in public passageway and space guaranteeing the feasibility of hotel in business. The hotel will accommodate about 500 guestrooms in the lower part of building and offer 250 high-class apartment-style guestrooms in the upper part of building. The landscape platform at the top floor will make the hotel become the landmark among Qingdao super star hotels. The design will full take advantage of ocean and seaport to create two completely different landscapes for the guestrooms. Hotel facilities will include platform, retailing, conference facilities, piazza, catering and business.

酒店透视图
Hotel perspective

青岛小港湾超五星级酒店单体设计及海岸线规划

规划/建筑设计：ANS国际建筑设计与顾问有限公司
设计师：Joe Lau、袁德澋、吕钊
用地面积：154 000 m²
建筑面积：128 500 m²
绿化面积：38 500 m²
基地面积：26 449 m²
建筑层数：69层
建筑密度：56.6%
容积率：3.91
绿化率：25%

青岛小港湾位于胶州湾东岸，城市老港区的南部，青岛市西部老城区，与中港、火车站、中山路相邻，岸线长度约2.8km。规划重新梳理了海岸线结构，"一带，两心，两轴"，设计以码头景观带串联起广场圆形舞台区域与主体建筑的两心，滨水的城市海岸景观轴和栈桥景观轴联动着人、海、城之间的整体空间，立体而有序地激发出滨水区域的空间活力。

酒店设计考虑创造青岛的最高结构在220m以上的当代城市地标，使之成为青岛城市文明的新象征。主题酒店设计与区域海岸线规划融为一体，相得益彰。酒店综合体提供一系列的娱乐和商业设施，在公共通道和空间内，确保了酒店在商业上的可行性。酒店将包括大约500个客房并位于建筑大楼的下半部分，另有250个高档公寓客房位于建筑物上半部分，顶层景观平台将成为青岛高星级酒店里程碑式的地标。充分利用海洋和海港，使酒店的客房提供2个截然不同的景观取向。酒店设施将包括主席台、零售、会议设施及广场、餐饮和商业。

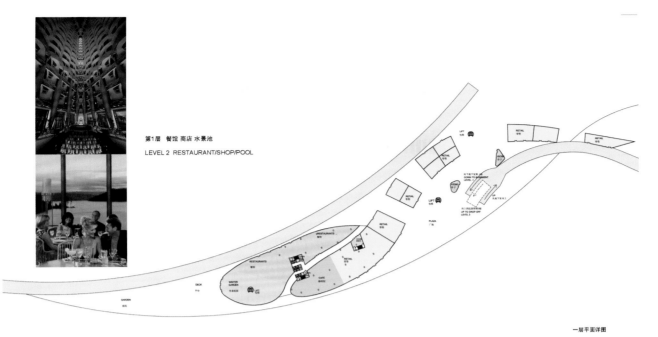

第1层 餐馆 商店 水景池
LEVEL 2 RESTAURANT/SHOP/POOL

一层平面详图

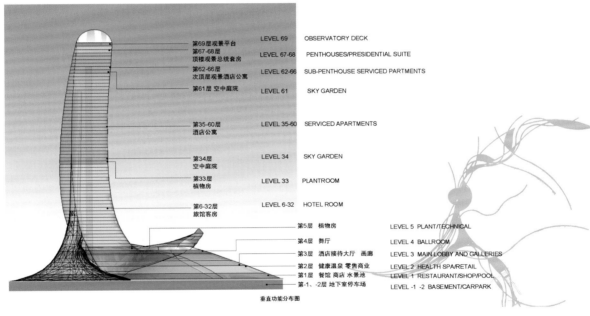

第69层观景平台	LEVEL 69 OBSERVATORY DECK
第67-68层 顶楼观景总统套房	LEVEL 67-68 PENTHOUSES/PRESIDENTIAL SUITE
第62-66层 次顶层观景酒店公寓	LEVEL 62-66 SUB-PENTHOUSE SERVICED PARTMENTS
第61层 空中庭院	LEVEL 61 SKY GARDEN
第35-60层 酒店公寓	LEVEL 35-60 SERVICED APARTMENTS
第34层 空中庭院	LEVEL 34 SKY GARDEN
第33层 植物房	LEVEL 33 PLANTROOM
第6-32层 旅馆客房	LEVEL 6-32 HOTEL ROOM
第5层 植物房	LEVEL 5 PLANT/TECHNICAL
第4层 舞厅	LEVEL 4 BALLROOM
第3层 酒店接待大厅 画廊	LEVEL 3 MAIN LOBBY AND GALLERIES
第2层 健康温泉 零售商业	LEVEL 2 HEALTH SPA/RETAIL
第1层 餐馆 商店 水景池	LEVEL 1 RESTAURANT/SHOP/POOL
第-1、-2层 地下室停车场	LEVEL -1 -2 BASEMENT/CARPARK

垂直功能分布图

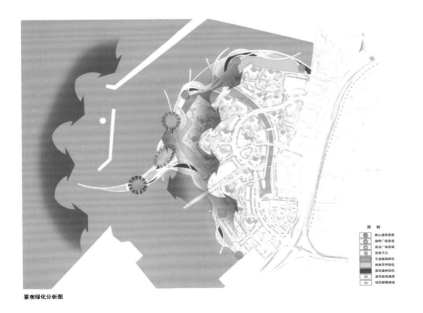

景观绿化分析图

6. 景观绿化分析

景观绿化以核心建筑景观为中心演绎广场、商业休闲分布置与灵动的滨海水湾，生态绿化、休闲草坪、都市森林立体而巧妙的构筑起了丰富的景观带，使得人在游玩赏海的同时还能体味绿色。蓝与绿在美丽的小港湾交融生长。

Urban Planning of Ecological Development and Regulation Section in Urban Section of Hutuo River, ShijiazhuangCity

Location: Shijiazhuang city, Hubei province
Base Area: 55km^2
Core Area: 12 km^2
Planned and Designed: ASPECTBJ Landscape Design Consulting Co., Ltd (Beijing)
Cooperative Design: German ISA International Design Group
Main Designer: Lilun, Zhang Yajin and Wangxin

Ecological Planning

In the ecological planning, riverway and northern shore of Hutuo River are mainly planned for ecological conservation to form a large riverside ecological wetland greenbelt. Cha River integrating existing park-style bank line will attract abundant urban life in the future forming the bank line combining city function and water tourism function; Xiaoqing River, as a small canal at present, is suggested to relate with Cha River in the south to form another channel for drainage. At the intersection of Cha River and Hutuo River, bank line is extended to serve for flood storage and provide diverse water landscape.

Industry Orientation

Basic on ecological construction, main supporting industry includes: enterprise zone for conference and exhibition, ecological travel, education, scientific research and original industry, and excellent residence. As the supplement of urban function, the planning will contribute to the increase of regional and urban economy through the central functional zone for international conference and exhibition. The area will become the important ecological tourism area of Shijiazhuang and the tourism basic facility and service center of Zhengding city by the construction of new zone for leisure, outing and tourist. The planning with this as basic includes the diverse urban function structure containing city-style and ecological high-class residence.

Landscape Planning

Ecological network, water system and street are applied to the design of visual corridor to relate spaces of different layers together while main visual corridors point to landmark public riverside space showing their excellent guiding property. Western side gives predominance to natural ecological landscape and eastern side tends to show the new looks of city. Different urban interfaces are designed basic on the spaces of different properties to pop out the characteristic of the area and create beautiful urban landscape.

The water area near the intersection of Cha River and Hutuo River is expanded to a lake forming the center of the whole eastern section and displaying the feature of "northern water". In the middle of the lake designs a landscape island with a 5-star hotel on to become the landmark of the whole central area and another landscape spot besides the riverside space adding the finishing point to the whole area. Main function clusters are arranged around central lake area to promote the value of land and create a beautiful urban space.

总体面图

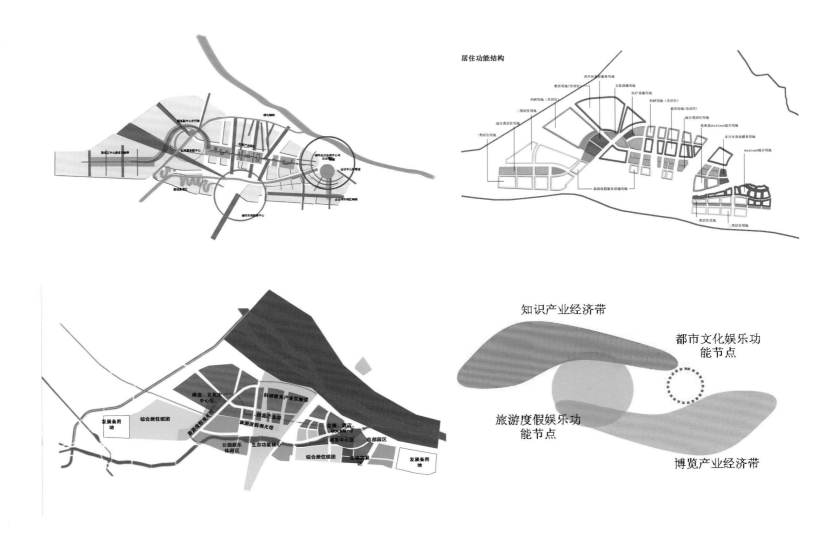

石家庄市滹沱河市区段生态开发整治区城市设计

项目地址：河北省石家庄市
基地总面积：55km²
城市设计核心区：12km²
规划设计：北京澳斯派克景观规划设计顾问有限公司
合作设计：德国ISA国际设计集团
主要设计人员：李伦、张亚津、王鑫

生态规划

生态规划中，滹沱河道及其北岸以生态涵养为主，形成大型沿河生态湿地绿带。汊河结合现有的公园型岸线，未来将吸引大量城市生活，形成都市型功能与水上旅游功能结合型岸线；小青河现状为小型运河，建议向南与汊河联系，形成汊河排洪的另一渠道。汊河与滹沱河汇集处，岸线有所拓展，一方面提供蓄洪水面，一方面提供更加多样化的水体景观。

产业定位

以生态建设为前提，强调主体支撑产业为：会议会展的企业园区、生态旅游、教育科研与创意产业、优质居住。作为城市功能的补充，通过国际会议、会展中心功能区将带来地区、城市发展的经济增长点。通过新休闲、郊游、对外旅游区的建设，成为石家庄重要的生态旅游活动区以及正定古城的旅游基础设施服务中心。以此为基础规划包括都市型与生态型高品质居住在内的多样化城市功能结构。

景观规划

利用生态网络、水系、街道等设计视线通廊，将不同层次的空间串联在一起，而主要的视线通廊均指向标志性的公共滨水空间，有很好的引导性。西侧以自然生态景观为主，东侧则倾向体现城市新貌的景观特点。结合不同性质的空间设计了不同的城市界面，突出本地区的特色，创造优美的城市景观。

在汊河临近滹沱河交汇处将水面适当放大成湖，形成整个东部片区的中心，凸显"北水"的特色。在湖中设计景观岛，岛上建设五星级酒店，成为整个中心区的地标和滨水空间的对景，起到画龙点睛的作用。主要功能组团均围绕中心湖区布置，提升土地价值，创造优美的城市空间。

Xuhui Riverside Public Open Space

Planned and Designed by: Shanghai Dblant Engineering Design and Consulting Co., Ltd

Fenglin Land
Land Area: 30 306.2m²
Floor Area: 35 478.6m²
Floor Area Ratio: 0.274
Building Density: 20.70%
Greening Ratio: 48.32%
Parking Space: 3711

Nanpu Station Land
Land Area: 14147.9m²
Floor Area: 9237.9m²
Floor Area Ratio: 0.25
Building Density: 16.66%
Greening Ratio: 46%
Parking Space: 101

In order to coordinate with the environment construction for Shanghai EXPO 2010, it has formally launched the "comprehensive development plan for banks areas of Huangpu River" including the world EXPO site affairs. Object of this architectural design is located in the eastern section of public open space (phase 1) Binjiang Xuhui District, which is also an important supporting node among the matched buildings here. The plots are triangle site on Fenglin Road and Nanpu block by railway station.

Design Concept
Reservation and reform: the base has been bearing witness to the history of industrial development in Shanghai for about one hundred years, representing the development of an era. To retain some one old building or structure is not just to keep the ones, but to preserve a memory belonging to the groups in an era by this way.

Experience
Just preservation is not enough, the design is to enable people

to experience a kind of wonderful memory, a type of dynamic change and a high quality of life style by means of old buildings preservation, structures' reform, the old and new architectures' amalgamation, vibrant place for landscape design, as well as the unique journey of industrial experience.

Overall layout
A complete building layout is applied on Fenglin plot, in contrast with Nanpu is taking the dismembered way. Although the two plots have totally different layout, both two are on their own suitable way.

Landscape system
Landscape system may emphasis on the better combination among the land and buildings. It develops a landscape system with features of geometry physique emphasized, simplicity and lordliness by deriving from a hexagonal motif.

Vertical System
Multilayered squares in sinking style can be set at the right sites using the level tolerance of original bases with the planned urban roads, flood proof walls with the wharfs on the premise of keeping original terrain, which would lead the main flow of people from city interface to the chief buildings in base step by step. The existing structures are also to be used for sky walk settings, connecting buildings at different levels together.

The amount of excavation for basement is to reduce as far as possible by taking advantage of the height difference in Fenglin plot. The two plots are tying best to get earthwork balance inside. The drainage devices are set to solve the issue of storm water discharge on the bases which are lower than local urban road level.

Architectural style
Design viewpoint: structures should be "appropriate" and "humble" in such an open space in Binjiang, no seeking for large-scale modeling or exaggerated way to express itself. Personality should be reflected in the most appropriate use of terrain, the perfect combination of old and new architecture and the wonderful space constructing.

The most exciting part of the original structures shall be kept, such as the pretty frame forms, sloping roofs, height differences with the surrounding land and so on. New functions are to be put into the old buildings inside, modern materials are used in the new parts and the old structures are given prominence as well. In this way, essence of the old buildings is kept and the unique personality of new buildings is displayed.

New architectural design comes to forge the image of humility and individuality through the physical characteristics, the new type material, contrast and integration with the old buildings.

枫林地块总平面图

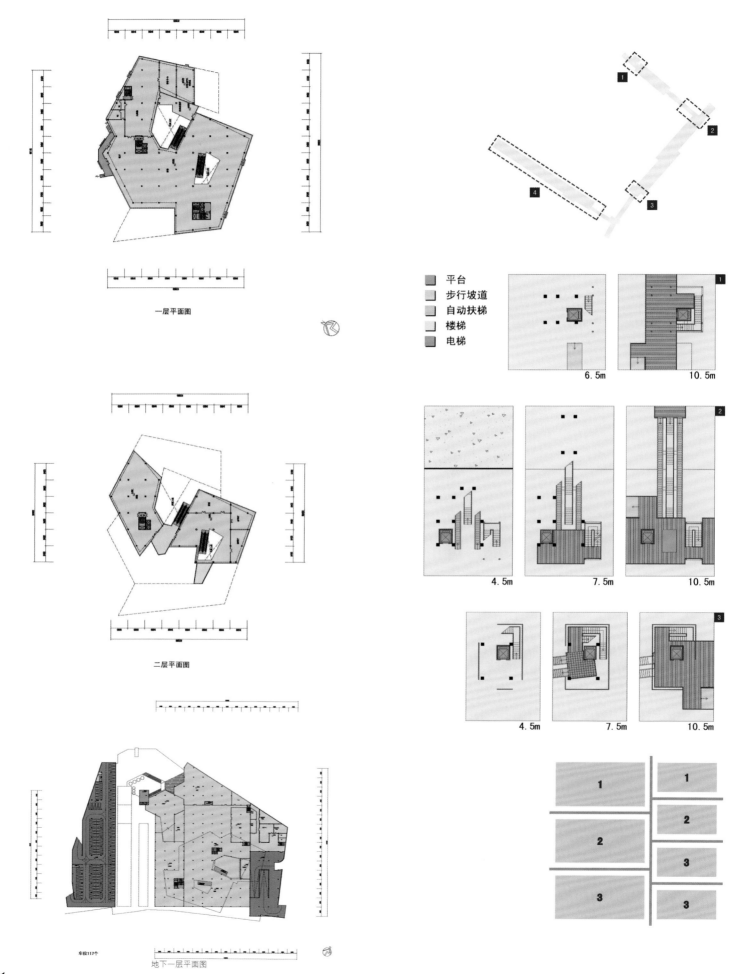

一层平面图

二层平面图

地下一层平面图
车位117个

平台
步行坡道
自动扶梯
楼梯
电梯

徐汇滨江公共开放空间

规划/建筑设计：上海都林工程设计咨询有限公司

枫林地块
用地面积：30 306.2m²
总建筑面积：35 478.6m²
容积率：0.274
建筑密度：20.70%
绿化率：48.32%
停车位：371l

南浦站地块
用地面积：14 147.9m²
总建筑面积：9237.9m²
容积率：0.25
建筑密度：16.66%
绿化率：46%
停车位：101

为配合2010年上海世博会环境建设，上海市正式启动包括世博会场在内的"黄浦江两岸综合开发计划"。本次建筑设计的对象位于徐汇滨江公共开放空间（一期）东段，为公共开放空间内的配套建筑中的重要结点，分别为枫林路三角地块和铁路南浦站地块。

设计理念

保留与改造：基地承载着同一段见证上海工业百年发展的历史，代表着一个时代的发展。保留基地内某栋旧建筑、某个构筑物，并不只是为了保存某个个体，而是希望通过这样一种方式来保存一种记忆，属于一个时代的群体记忆。

体验：仅仅保留是不够的，设计通过旧建筑的保留、构筑物的改造、新旧建筑的融合、富有活力的场所景观设计，以及独特的工业体验之旅，让人们体验一种美好的记忆，体验一种有活力的变迁，体验一种高品质的生活格调。

总体布局

枫林地块选择完整的建筑布局，南浦站地块化整为零，两个地块所采用的布局方式有所不同，但却是适合各自基地的方式。

景观系统

景观系统强调与地块及建筑更好的结合，以六边形为母题，衍变而成一个既强调几何形体、又简洁大气的景观系统。

竖向系统

尊重基地原有地形，利用原有基地与规划城市道路、防汛墙以及码头的标高差，在合适的位置设置多层次的下沉式广场，将主要人流一步步从城市界面引导入基地内主要建筑。利用基地原有构筑物，设置空中步道，将不同层面上的建筑联系到一起。

枫林地块利用高差，尽可能将地下室开挖土方减少。两个地块尽量做到内部土方平衡。到在局部低于城市道路标高的基地内，设置强排水装置，解决雨水排泄问题。

建筑风格

设计观点：在这样一个滨江开放空间内，建筑应该是"适宜"的、"谦逊"的，不寻求通过大尺度或者夸张的造型来表达自我。个性应该体现在对地形的最适宜利用、新旧建筑的完美结合、丰富的空间营造上。

保留原有建筑最为精彩的部分，比如漂亮的结构形式、坡屋顶、与周边地块的高差，将新的使用功能置于旧建筑内部，新的部分通过现代的材料、局部突出旧建筑的方式，既保留旧建筑的精华部分，又彰显新建筑的独特个性。

新建筑设计通过有特色的形体、新型材料、与旧建筑的对比与融合，来塑造一种既谦逊又不失个性的建筑形象。

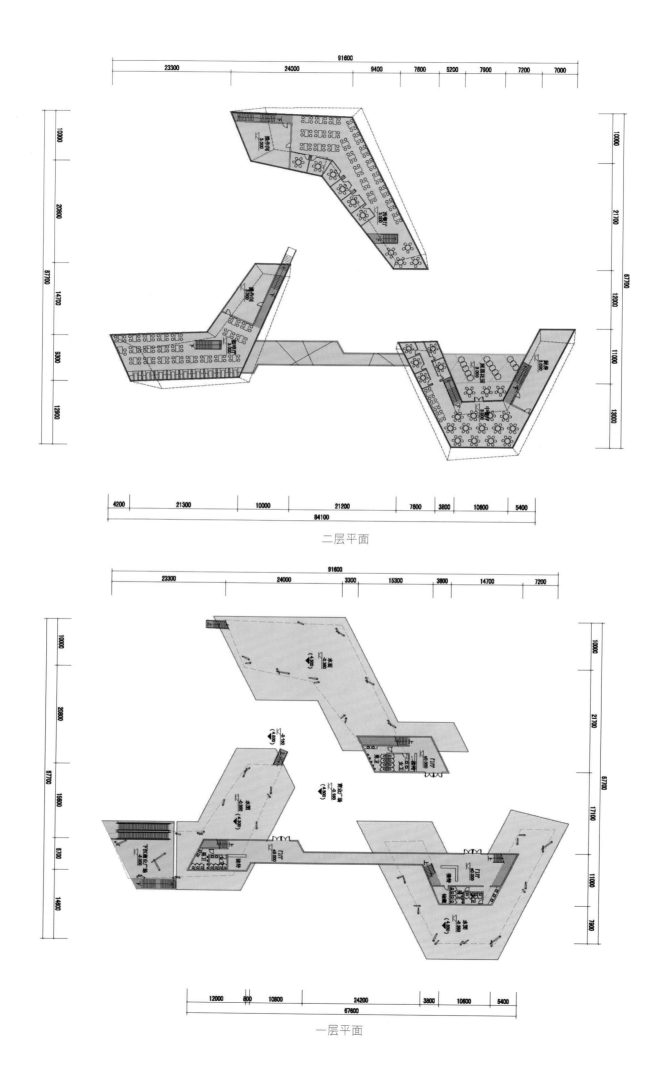

二层平面

一层平面

264

Hangzhou "Riverside City" Concept Plan

Architectural Designer: Dblant Design International

The planning area for Hangzhou "Riverside City" starts from the riverside Xiaoshan District at the east to Zhejiang-Jiangxi Railway-Puyan Road at the south and to Qiantang River at the northwest, covering the total area of 37.7 square kilometers. The High-tech Zone (Binjiang District) is along the river and clings to the bridge with convenient traffic and unique geographical location.

Idea+Tools
Through rational use of sustainable resources, create a fully functional, beautiful and harmonious community environment, where the energy use and refuse production is minimized and the resource conservation and waste recovery is maximized.

The sustainable community environment under the guidance of a balanced way will lead to social prosperity, economic growth and environmental optimization. Focus on developing a healthy lifestyle and the concept of outdoor activities, as well as the creation of an environment able to enrich and activate individual and family life.

Therefore, in order to create such a community we must provide appropriate services and facilities, such as education, health, community center and open space for activities. These elements are the core contents to create a successful surrounding environment.

For this purpose, it is necessary to create and implement a "municipal tool box" with design regulations and functional measures.

Planning structure
The plan focuses on improving urban functions and centers at key developing regions, through the connection of major city roads and Riverside Avenue, developing the surrounding area in a radiating manner. The key developing regions are as follows: Corporate Headquarter Zone, E-commerce Zone, Creative Industrial Park, as well as Prospective Business District.

Transport system
Combined with the actual situation in Hangzhou, the plan increases two cross-river tunnels and two cross-river subways to make more convenient cross-river transport between Binjiang District and the main city of Hangzhou and better interactive development with other districts of Hangzhou City.

Landscape system
Binjiang District has great location landscape advantages, so systematic planning and design is very necessary for urban green landscape system. The plan takes water as the landscape focus, focusing on planning and building riverside scenic belt, while making full exploitation of the waterside landscape in the reconstruction base, continuing road greening so that the green landscape in the entire Binjiang District forms a complete and effective system. The Riverside City in future will take on a garden city scenario "with the river around the city, the city in the green, people in the scenery and the scenery in the water".

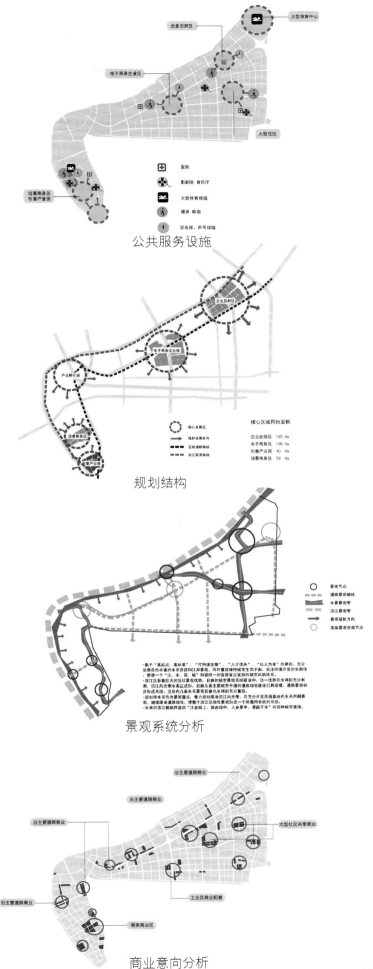

公共服务设施

规划结构

景观系统分析

商业意向分析

杭州"滨江新城"概念规划

建筑设计：上海都林工程设计咨询有限公司

杭州"滨江新城"规划范围东起滨江萧山区界，南至浙赣铁路—浦沿路，西北至钱塘江，规划总用地37.7km²。高新（滨江）区沿江依桥，交通便捷，地理位置得天独厚。

理念+工具

通过合理利用可持续性资源，创造一个功能齐全的、美丽而又和谐的社区环境。在这个社区，能源利用和垃圾产量得到最小化，资源节约和垃圾回收得到最大化。

可持续性的社区环境在平衡的方法引导下可使社会繁荣，经济增长，以及环境优化。注重培养健康的生活方式和室外活动的理念以及注重创造一个可丰富活跃个人和家庭生活的环境。

为创造这样的社区需提供相应的服务设施，如教育，医疗卫生，社区中心和开放的活动空间等。这些要素是创造一个成功的周边环境的核心内容。

为此目标，需要创造和实施一个拥有设计法规和功能措施的"市政工具盒"。

规划结构

规划以完善城市功能为重点，以重点发展区域为核心，通过主要城市干道及滨江大道的联系，辐射发展周边区域。规划重点发展区域分别为：企业总部区、电子商务区、创意产业园以及远景商务区。

交通系统

结合杭州的实际情况，规划增加跨江隧道两条、跨江地铁两条，使滨江区与杭州主城的跨江交通更为便捷，更好地与杭州其他城区互动发展。

景观系统

滨江区有着巨大的区位景观优势，系统的梳理和规划设计对城区的绿化景观系统是十分必要的。规划将水系作为景观重点，着力规划建设沿江风光带，并充分开发改造基地内水系两侧景观，继续建设道路绿化，使整个滨江区绿化景观形成一个完整而有效的系统。未来的滨江新城将呈现"江在城上，城在绿中，人在景中，景融于水"的园林城市意境。

杭州"滨江新城"概念规划

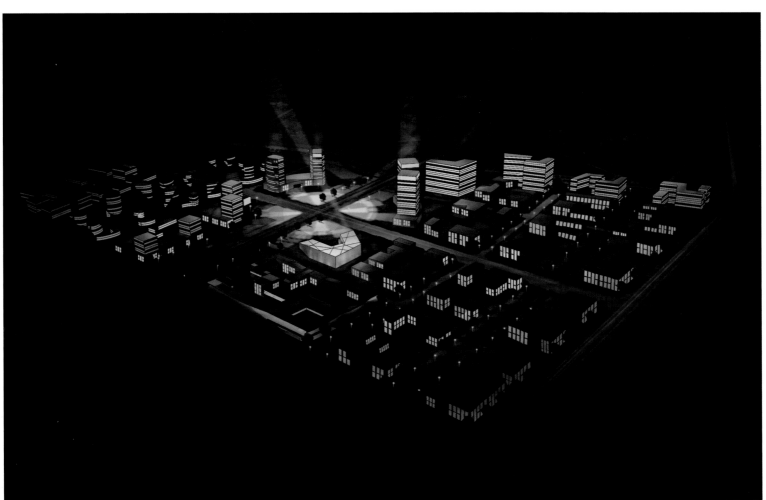

Guigang Business Zone on No.GB-16 Land, Gangbei New District

Planned and Designed by: Guangzhou Weilun Construction Design Consulting Co., Ltd
Designer: Fan Wenfeng, Su Weifeng, Tanjian

Land Area: 98 252.6 m^2
Floor Area: 179 508 m^2
Parking Space: 780
Construction Density: 27.34%
Floor Area Ratio: 1.80
Greening Ratio: 56.3%

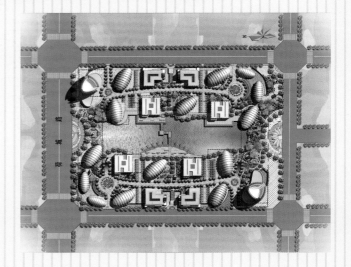

This design is featured by the concept of "human-oriented" aiming to construct an ecological environment and create a well-arranged, convenient, green and cultural office zone. The design will pay attention to the ecological environment inside the zone, reasonably distribute and use various resources and fully embody the concept of sustainable development. The scheme will perfect various basic matching facilities and determine various reasonable control standards. High starting point, high standard and reasonable consideration of the relation between administrative center and cultural center will be also included. Designers are to extend the historical and cultural context and create a landmark business zone which is with abundant cultural deposits inside and a brand-new appearance outside. The design will materialize the principle of participating and controlling to improve the flexibility of the layout in the district. While guaranteeing the wholeness of ecological system, designers also enable the concrete operation to be carried out by stages to offer the convenience of independent use.

贵港市港北新区GB-16地块办公商务区

规划设计：广州市纬纶建筑设计顾问有限公司
设计师：范文峰，苏伟锋，谭剑

占用地面积：98 252.6 m²
总建筑面积：179 508 m²
停车位：780
建筑密度：27.34%
容积率：1.80
绿化率：56.3%

设计方案贯彻"以人为本"的思想，以建设生态型环境为规划目标，创造一个布局合理，交通便捷，绿意盎然，生活方便，富有文化韵味的办公区。注重规划区的生态环境，合理分配和使用各项资源，全面体现可持续发展思想。完善各项基础配套设施，合理确定各项控制指标。坚持高起点，高标准的规划，合理考虑行政中心和文化中心的关系。延续历史文脉，创造一个内在悠久文化底蕴、外表全新的标志性的办公区。规划设计体现公众参与开放空间和可操作性的原则，加强该地区规划设计的灵活性。在保证整体生态系统的前提下使具体操作便于分期实施，做到独立使用的方便性。

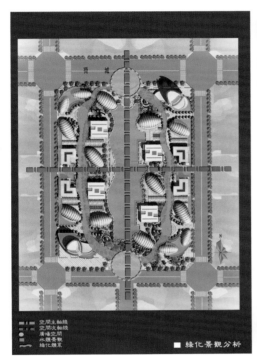
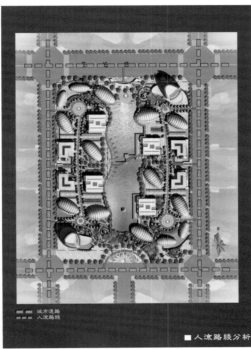
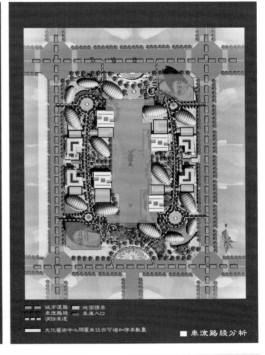

Aachen Institute of Technology (Melaten Campus) Planning

Location: German Aachen
Designed by: German S.I.C-Engineering Consulting Co., Ltd
Designer: Jan Gutermuth
Floor Area: 259 964 m²

The Aachen Information Science and Technology Park is a business card of Aachen Institute of Technology. Research, teaching, development and life are the basic construction contents of the Park. The design gives the existing Aachen (Melaten Park) a structured system so that its single parts connect into a larger whole, so its total investment will cover the entire scope of the campus.

Melaten Park has currently been influenced by its wide range of landscapes and green spaces, which will thus be also adopted by the design program: to extend out a green strip from the central green space and reinforce the green strip through new buildings to make it throughout Melaten Park. The green strip will connect each old and new area into a highly practical and living whole area.

Consideration of urban space
The distribution of the green strip is in accordance with the relevant design and planning, developing to the north based on the continuously changing and extending terrain until to replace the landscape facilities of the inner park of Aachen as well as connect the Information Science and Technology Park.

Pearl
The green strip is arranged like a string of pearl necklace as another focus of public life opposed to the campus. It has square forms of different quality, which is composed of different areas on different cut-off point in the green strip.

Green strip –Solitaire
The reconstructed Solitaire-style construction determines the building direction of the green strip. She is the architectural center of the campus life.

Building cluster – Area
The campus is composed of the old park and the new park. Most importantly the new cluster must be within the designed scope. Six clusters composed of type 2 and 3 enclose a park-like central square, connected with the green strip three-dimensionally by an existing green tongue way. The exterior surface of all the buildings faces the square (with no negative frontage).

Cluster
The building cluster is based on a module system composed of only three modules. The building mode of the module is suitable for the requirement to meet the growing number of users in the building cluster.

The urban planning has established a framework to enable the cluster and the module in unique and flexible combination.

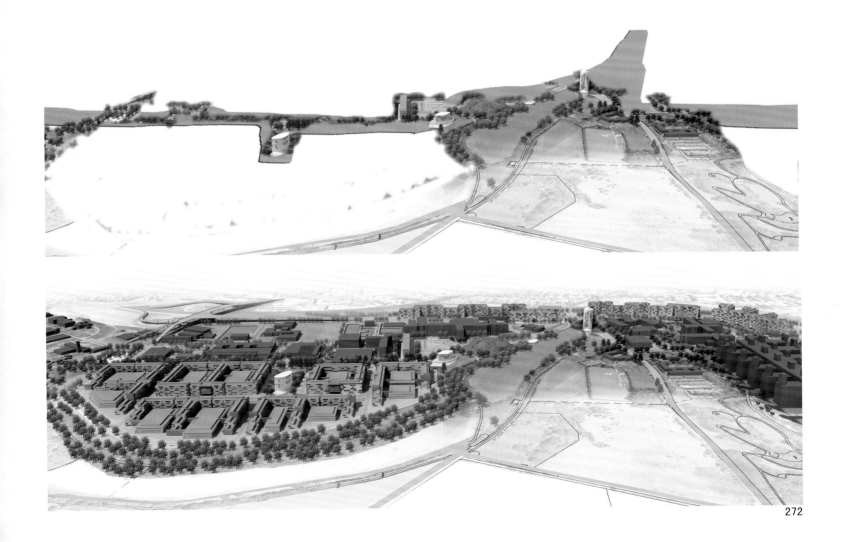

平面图 平面图 平面图

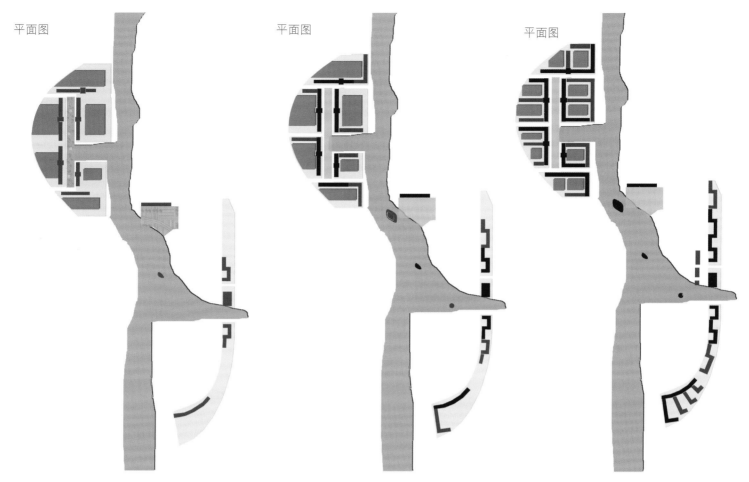
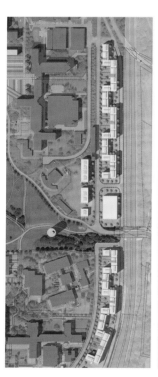

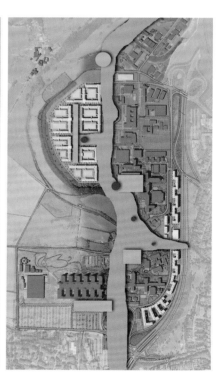

Westansicht Mustercluster / Quartierplatz

立面图

立面图

立面图

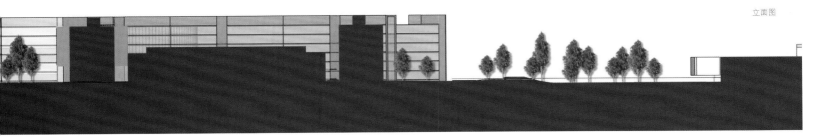

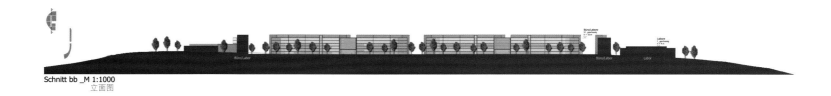
Schnitt bb _M 1:1000
立面图

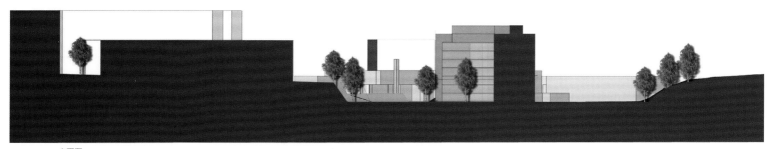
立面图

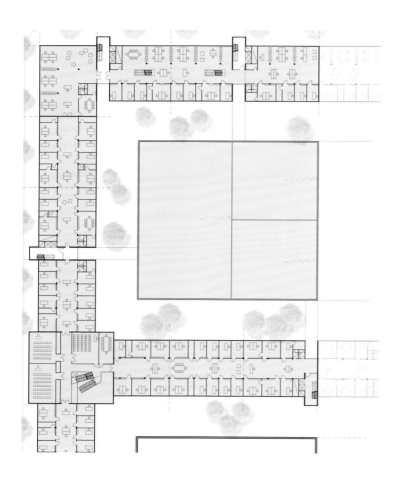

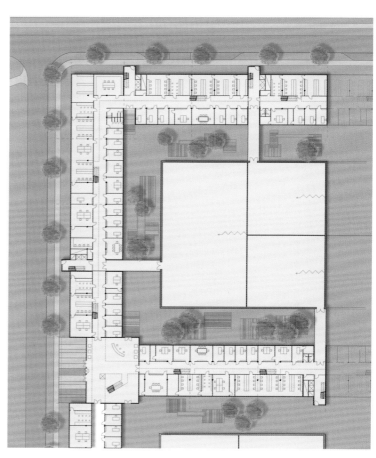

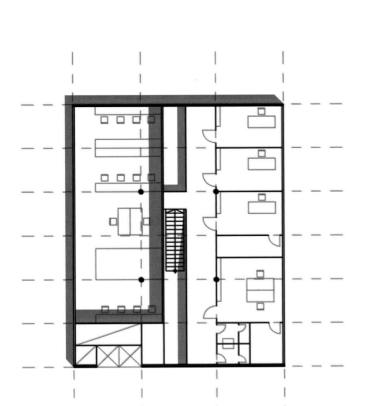

平面图

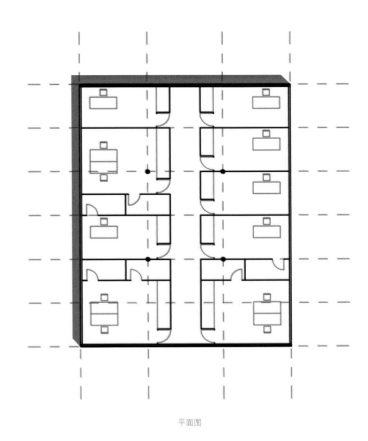

平面图

亚琛工业大学校园(Melaten园区)规划

项目地点：德国亚琛
设计：德国S.I.C.-工程咨询责任有限公司
设计师：杨·古特木特(Jan Gutermuth)
建筑面积：259 964 m²

亚琛的信息科技校园区是亚琛工业大学的一张名片。研究、教学、发展、生活是这个信息科技园区的基础建设内容。方案设计给现有的亚琛工业大学(Melaten园区)一个结构体系，使它的单一个体连接成一个大的整体，因此它的总投资将覆盖整个校园的范围。

Melaten园区目前已经被它的多种多样的景观以及绿色空间的塑造所影响。这将被方案继续所采用，即从目前已拥有的中心绿地继续延伸出一条带状绿化，并将此带状绿化通过新建筑的音符来加强，使之贯穿整个Melaten校园。这条带状绿地把每一个旧的和新的区域连接成一个生活性及实用性极强的整体区域。

城市空间的考虑

这条带状绿地的分布，是根据相关的设计规划，按照不断变化延伸的地势向北发展，直到可以取代亚琛工大内部园区的景观设施，同时也把信息科技园区连接到一起。

珍珠

在这条带状绿地的排布上，就如同一串珍珠项链，相对于校园是另外一个公共生活的焦点。它有着不同品质的广场形式，这是由这条带状绿地上各个切点同不同的区域共同组成的。

带状绿地——接龙

重新规整功能的接龙式建造确定了这条带状绿地的建筑走向。她是校园生活的建筑中心。

建筑群——区域

这个校园是由原来旧的和新的园区组成的。最重要的是，新的集合区域要在设计规划的区域内。由类型2和3组成的六个集合区围合成一个公园式的中心广场，通过一个已有的绿色舌道与带状绿地立体的联系到一起。所有的集合群的外墙面都面向于广场（没有不利的朝向）。

集合群

这个集合群是基于在一个只有三个模块组成的一个模块体系上的。这个模块的建造方式适合于建筑集合群在满足不断增多的用户情况下的要求。

城市规划已确定了一个框架条件，使集合群能够十分灵活独特的与模块共同结合起来。

COMPREHENSIVE COMMUNITY

综合

Everbright International Center

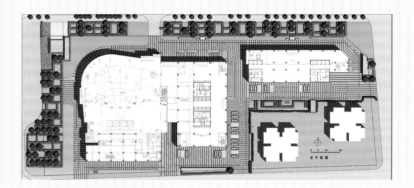

Location: the interchange between Fuchengmen North Street and PingAn Street, Xicheng district of Beijing.
Architecture Design: Zhongyuan International Engineering Design Institute
　　　　　　Chinese Academy of Sciences, Beijing Architecture Design Institute

Land Area: 22 000 m²
Floor Area: 160 000 m²
Frame: frame-shear wall structure

Everbright International Center is a comprehensive landmark business construction cluster located at the No.28, West Street, Ping An Li, Xicheng District, Beijing city. The location is the crossway between West Second Ring Road and PingAn Street and the Gongzhuang station of metro line 2 is just at the northwestern corner of the project. The total floor area is 160 thousand square meters and overall planning includes office building, business service, hotel and theatre.

The building 1 is designed for grade A office building with floor area 64917.06 square meters, standard storey height 3.8 meters, 20 stories in all, total altitude 82.6 meters and standard floor area 2429.11 square meters. Business services are offered from storey 1 to storey 3 with banks, business center and catering, etc.

一层平面图

二层平面图

光大国际中心

项目地址：北京西城区阜成门北大街与平安大街交汇处
建筑设计：中元国际工程设计研究院
　　　　　中科院北京建筑设计研究院

占地面积：22 000 ㎡
总建筑面积：160 000 ㎡
建筑结构：框架剪力墙

光大国际中心是一个地标性综合商务建筑群，位于北京西城区平安里西大街28号。地处西二环路与平安大街交叉路口，地铁二号线车公庄站坐落项目西北角。总建筑面积16万平方米，整体规划有写字楼、商业、酒店、剧院等物业形式。

光大国际中心1号楼为甲级写字楼，建筑面积64 917.06㎡，标准层层高为3.8m，20层，总高度为82.6m，标准层面积为2429.11㎡，其一至三层为商业，特设银行、商务中心、餐饮等。采用超5A智能

280

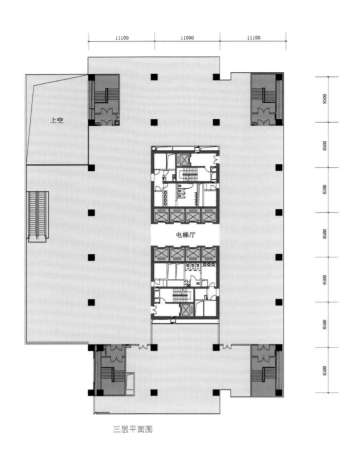
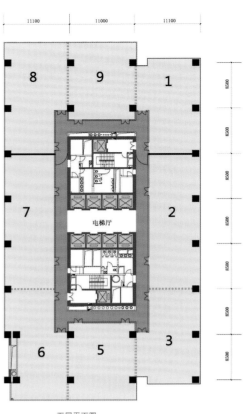

三层平面图　　　　　　　　　　　　五层平面图

办公标准。国际知名楼控系统智能控制大厦机电设备，强大的网络拓展能力，使大厦内不同的机电系统通过楼控系统组成一个集群，更安全、舒适及节能。电梯系统采用4/4电梯优化群控，10部日立电梯高速高效运行，提升办公效率。电信运营方面通过高端电信基础和增值服务，提供一站式"7X24小时"服务、定制化全业务电信解决方案。相邻于1号写字楼，国际四星级酒店入驻，构筑国际商务社交空间；同时，坐享区域人文标志中国国家级京剧院，亦商亦赏，与客户及商务伙伴于此共演永不落幕的商务大戏。

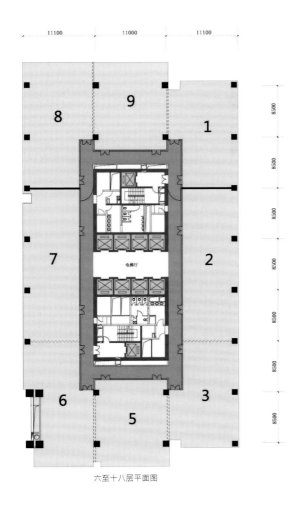

六至十八层平面图

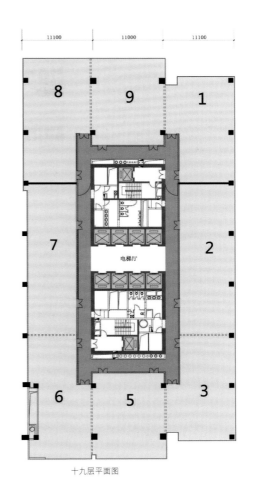

十九层平面图

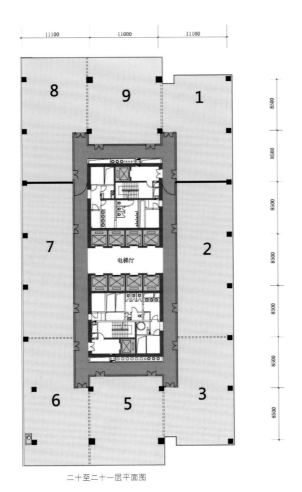

二十至二十一层平面图

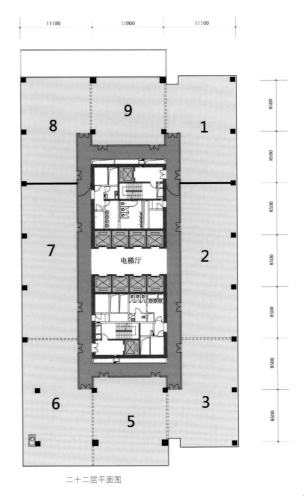

二十二层平面图

Shanghai Chunagzhan International Trade Center

Designed by: Huasen Architecture and Engineering Design Consultants Company Limited

Shanghai Chunagzhan International Trade Center is invested and developed by the Pearl River Group in Zhao Lane, Qingpu District, Shanghai with the area of 419,000 square meters. It sits between the A9 highway in the south and Huqingping Road in the north, opposite to OUTLETS Brands Square. Total construction area is planned to be 976,000 square meters. There is a commercial area in the east of project, principally based on main large-scale commercial and exhibition center, a mixed-use zone for leisure entertainment, hotels, hotel-style apartments and apartment-style offices. It will become the total radiation of Shanghai as far as its surrounding region of collecting and distributing large-scale commercial and leisure resort upon completion. The five-star hotel takes an area of 51.06 thousand square meters. In addition to 500 rooms, there is matching facilities like hotel-style apartments of 48,006 square meters, bath and entertainment center of 27,163 square meters. The overall planning and design is currently being submitted for approval.

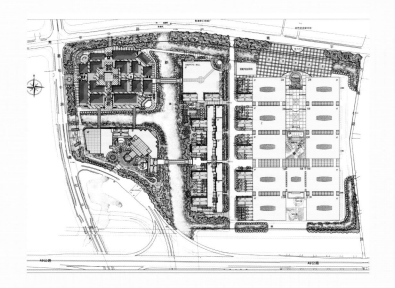

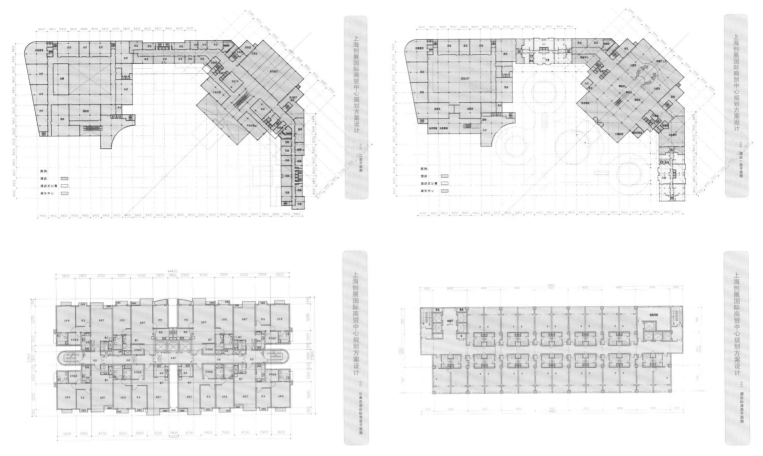

上海创展国际商贸中心

建筑设计：华森建筑与工程设计顾问有限公司

上海创展国际商贸中心是珠江集团在上海市青浦区赵巷投资开发，用地面积41.9万m^2，南临A9高速公路，北临沪青平公路，对面为OUTLETS品牌广场，拟规划总建筑面积97.6万m^2，项目的东区为商贸区，以大型商贸和展示中心为主；西区为综合区，作为休闲娱乐、酒店、酒店式公寓和公寓式办公，建成后将成为辐射全上海及其周边地区的大型商贸集散及休闲度假中心。其中五星级酒店建筑面积51060m^2，除有500间客房外，还配套设有酒店式公寓48006m^2与洗浴、娱乐中心27163m^2。目前总体规划设计正在报批中。

Taikoohui Comprehensive Development, Tianhequ, Guangzhou

Location: Tianhequ, Guangzhou
Site area: 49 000 m^2
G.F.A.: 446 000 m^2
Developer: Swire Properties Ltd

This mixed-use development is located in Tianhequ, Guangzhou, China and is composed of a 40-storey and a 28-storey office tower/boutique hotel, a 5-star Hotel and a Cultural Center. Each of these is located on a corner of the site, ensuring views for each building are not blocked and that there is privacy between the different buildings. The three towers are connected via a plaza which includes a convention center, commercial spaces, and a shopping mall as well as a major transit interchange. The main entrance to the entire venue is through the lobby in a glass cube which is connected to the retail plaza.

广州天河太古汇综合发展项目

地点：广州天河区
用地面积：49 000m²
总建筑面积：446 000m²
开发商：太古汇地产有限公司
建筑设计公司：梁黄顾设计顾问(深圳)有限公司

设计说明：广州天河太古汇综合发展项目位于天河路交汇处的西北方向,南临天河路，东临天河东路，北临规划中的N1路，紧靠项目南面的是兴建中的地铁3号线石牌桥站，西侧是地铁1号线体育中心铁。针对这一情况，设计者计划利用地下隧道否将地铁站与项目的地下二层相连。在整个项目布局上也进行了精心的分析，一座40层高的办公楼和一座28层高的办公楼，精品酒店分别坐落于用地的东南和西南角；五星级酒店坐落于用地的东北面；文化中心刚位于基地的西北角；一座四层的商场裙楼将以上各部分联系一起；整个项目还包括地下三层车库，地下四层的装卸货及设备层。

Shenzhen Line 4 (Phase 2) Property Development

Location: Longhua Extension Area, Shenzhen
Site area: 240 029 m²
G.F.A.: 626 880 m²
Developer: MTR Corporation Ltd

Line 4 (Phase 2) of the Shenzhen Metro is a 16 km long mass transit line with ten stations and associated property developments along it. Located in the Longhua Extension Area, this project involves the urban planning of four of the sites, all of which includes residential, commercial and institutional properties above or adjacent to the stations.

Under development by the MTR Corporation, Shenzhen Metro Line 4 Phase 2 will embody the comprehensive rail plus property model which has been successfully realized in Hong Kong. With the rapid growth of Shenzhen, this urban planning scheme will accommodate the future demand of this part of the city, enabling it to grow sustainably by providing mass transit to these areas.

Each station along the railway will be a free-standing low-rise structure containing clearly articulated retail layers that allow direct access to and from the concourse. A mixture of high and mid-rise residential developments will be set back from the station, introducing a green buffer zone through the development. This will clearly mark circulation routes and open up view corridors, as well as creating outdoor recreational space. Each development will also include serviced apartments, hotels, clubhouses and kindergartens.

深圳市轨道交通四号线二期沿线物业发展

地点：深圳龙华二线扩展区
用地面积：240 029m²
总建筑面积：626 880m²
开发商：地铁有限公司
设计公司：梁黄顾设计顾问(深圳)有限公司

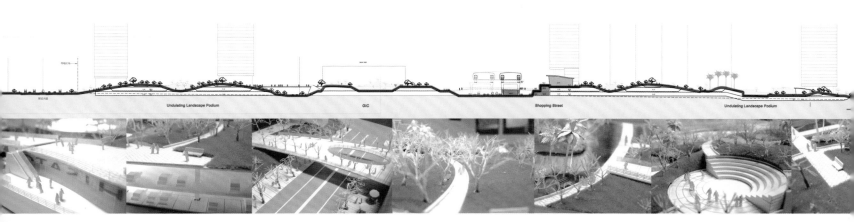

深圳地铁四号线二期全长16公里，包括十个车站及相连物业发展项目。项止坐落在地点优越的龙华二线扩展区，建筑设计涵盖四幅发展用地的城市规划，当中包括位于车站上盖和相邻的多种住宅、商业及公用物业项目。

这个铁铁路项由香港地铁公司发展，顺理成章采用香港行之有效的模式，即以完善的铁路系统加上物业发展经营。鉴于深圳市现正急速发展，龙华二线扩展的城市规划，采取了可持续发展的原则，再配合铁路运输和多元化的物业发展模式。

在设计的构思上，每一个车站均是一座显眼的独立建筑，除此之外每个车站都附设商场，并设有通道连接车站大堂，借此增加购物人流。在车站与四周建筑物的布局方面，刻意将层数较多的住宅项目放在距离车站较远的地方，好腾出空间来建设绿化带，树立清晰的行人指示牌、设立观景走廊及创造更多的休憩空间。此外尽量使区内服务式住宅、酒店、会所、幼儿园及其他商业设施一应俱全，使整个项目进一步融和当地社区的文化生活，使人们的生活、工作有一个无缝对接。

Comprehensive Development of Plot 9, Danang, Vietnam

Designed by: ANS International Architecture Design and Consulting Co., Ltd
Designer: Tony, Jeffry, Zhanghai

Land Area: 90 285 m²
Floor Area: 339 758 m²
Base Area: 90 285 m²
Number of Stories: 32
Building Density: 34.8
Floor Area Ratio: 3.76

A boulevard is designed to connect office building, exhibition center and 5-star hotel together serving for convenient traffic. The project includes a People's Piazza located behind business center and between office building and hotel offering a space for rest, sightseeing, catering and artistic works. The footpath of piazza integrates all functional blocks into one. High-rise office building becomes the landmark of the plot and is also the new symbol of luxurious apartment cluster in Danang. These building clusters are built along the river forming different angles in order to give the best sight to apartment buildings and 4-star hotel. Various dining-rooms, coffee shops and stores are set to decorate footpath beside river and form a dynamic and vivid scene.

越南DANANG 9号地块综合开发

设计：ANS国际建筑设计与顾问有限公司
设计师：TONY、JEFFRY、张海

用地面积：90 285 m^2
建筑面积：339 758 m^2
基地面积：90 285 m^2
建筑层数：32层
建筑密度：34.8
容积率：3.76

一条林荫大道连通办公大楼，会展中心和五星级酒店，交通便捷。项目中设有人民广场，位于商业中心后，办公大楼与酒店间，作为人们休憩、观景、餐饮及艺术作品聚集之地。人民广场中设有步行道，将所有主要功能块联系起来成为整体。高层办公大楼成为该地块的地标建筑，也是Danang豪华公寓楼群的新标志。这些楼群依河而建，面向河流各成角度，使得公寓楼与四星酒店都能获得最佳的山河景观。各类餐厅，咖啡馆，小店点缀于河岸步行道边，形成动态的具有活力的景观。

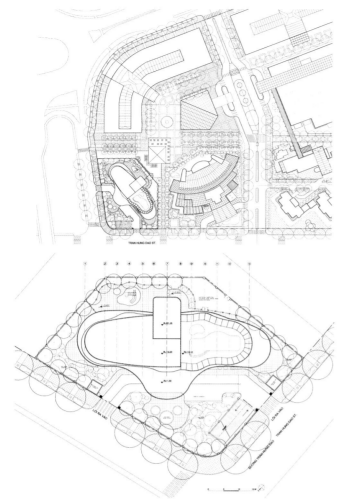
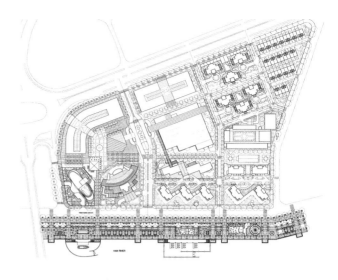
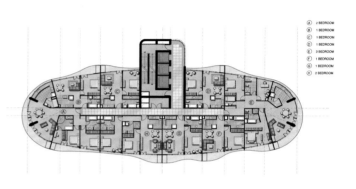

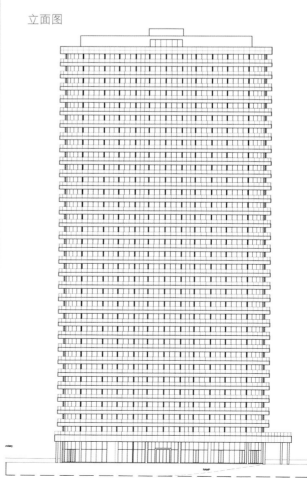

立面图

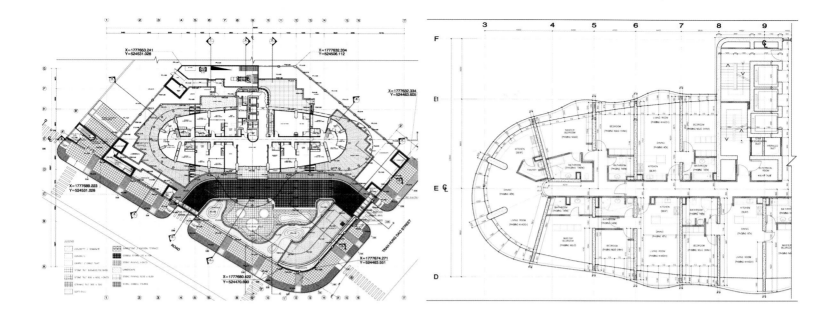

JingAn No.60 Lane

Location: Shanghai
Designed by architect: Zhang Qingyue, Dblant Desgin International Dongji Design

Land Area: 17 940 m²
Floor Area: 178 196 m²
Floor Area Ratio: 7.6
Building Density: 49.40 %
Greening Ratio: 20.04 %
Parking Space: 730

The foundation of this plan is located in the river bank of Suzhou, and a sculpture art park is planned to be built in the south of the plot, which proves a considerably superior geographical location. Under this circumstance, a Chinese-style modern art and culture center commercial area is proposed to be established, and it will create a good radiation effect to the entire plot as well as its neighborhood once it is built up.

This project takes the original city walls as the concept base. In the past, the city walls were built up for the purpose of defending the

invasion of the enemies from the sea; since the reform and opening-up policy was carried out, various types of architectures designed by Europe and America have been swarming into Shanghai as what they did in those days. We need a city wall likewise…

The concept of this plan is to joint five plots to be a single whole to form another culture and art commercial center within Jing'an Area.

The concept of business center is to reflect the composite business room constituted by associating the two business forms of the indoor large space of plot A and the outdoor large space of plot B so as to create a new active center.

Planning concept
1. According to the original city wall concept in Shanghai, five plots are conceptually combined (SOHO, office building, etc.).
A symbolic tower (a height of 250m) is proposed to be established at the plot A to be the new landmark of the west side of Huangpu River. The idea of SOHO modeling originated from the "sailing boat", a sea transport in the 19th century with an arc-shaped enclosure(the concept of city wall), the implied meaning of which is to become the Shanghai boat of world culture.
2. It adequately presents the energy-saving and environmental-friendly spirit in the current world. The modern energy-saving spirit and the determination of protecting the current situation of nature are connected in the form of 3-D green belt, which run through the whole foundation.
3. A business center specific to Shanghai is established. The plot A, which combines with the exit of Shanghai subway station, will bear a huge flow of people. A large space is built up at the exit of the subway station to enable people to consciously ignore the existence of subway as much as possible. Coming out from the subway, a large-sized business space will be the first object which catches one's eyes, making people to sensually ignore the subway.

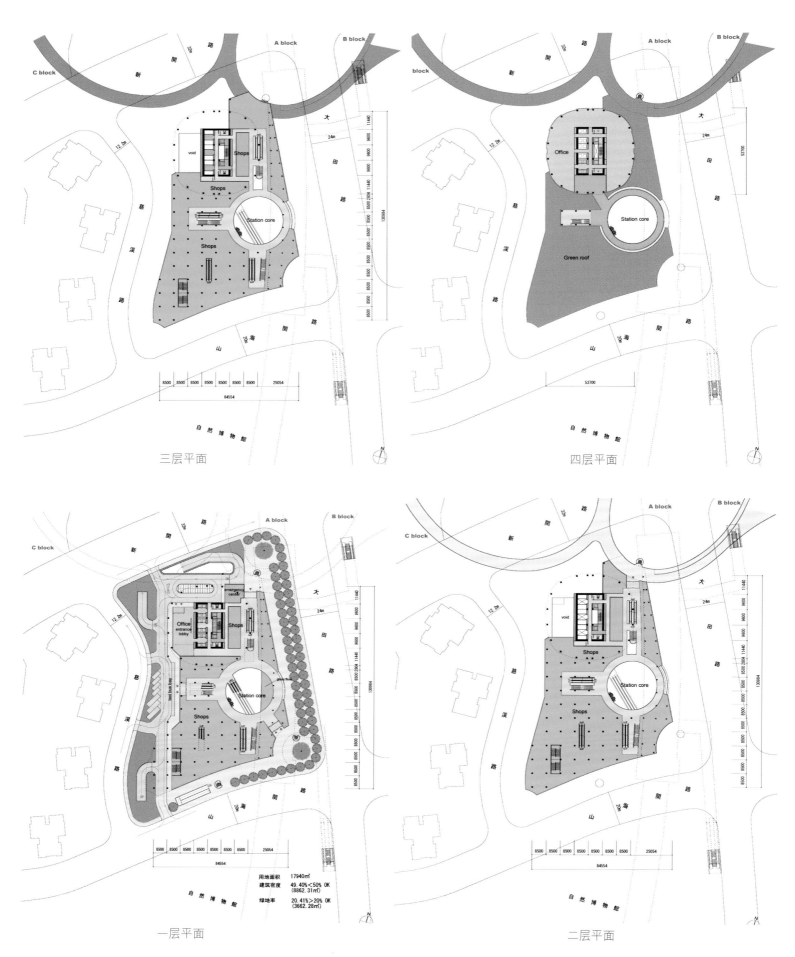

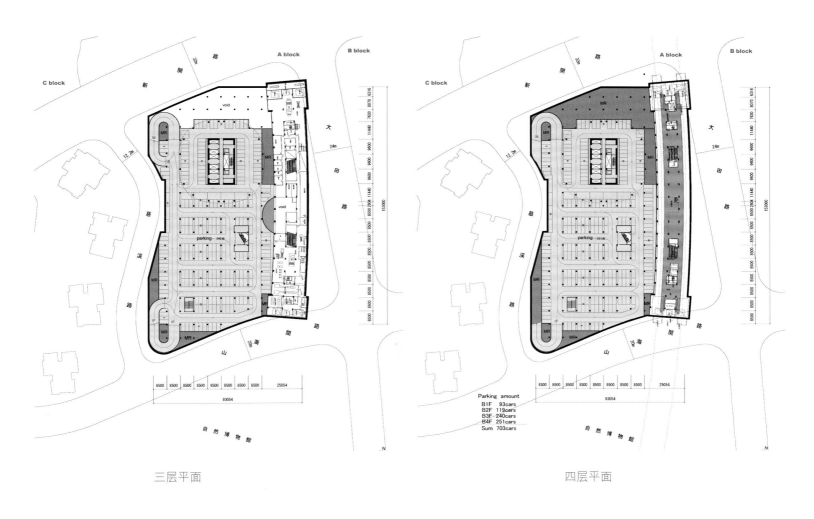

三层平面 　　　　　　　　　　　　　　　　四层平面

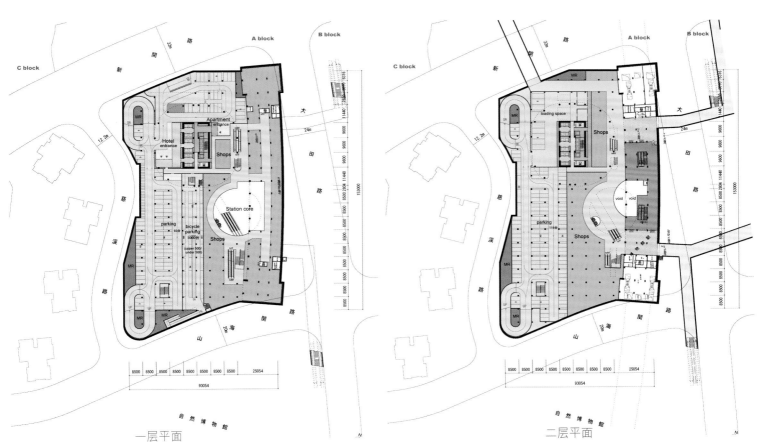

一层平面 　　　　　　　　　　　　　　　　二层平面

静安60#街坊

项目地址：上海
建筑设计：都林国际设计 张清嶽建筑师
　　　　　东急设计

用地面积：17 940 ㎡
总建筑面积：178 196 ㎡
容积率：7.6
建筑密度：49.40 %
绿地率：20.04 %
停车位：730

　　本方案基地位于苏州河畔，地块南侧规划建立一座雕塑艺术公园，地理位置相当优越。在此条件下，规划拟建一座中国式的，现代化的艺术文化中心商业小区。一旦建成将对整地块及地块周边地区形成良好的辐射作用。

　　本项目以原先的城墙作为概念的基础，以往建造城墙是为了防御敌人从海上入侵，自改革开放以来，欧美各种形式的建筑，如当年一般蜂拥侵入上海，我们同样也需要一座城墙……

　　本方案的理念是将五个地块联合成一体，成为静安区又一个文化艺术商业中心。

　　商业中心的理念是体现将A地块室内大空间与B地块室外大空间的商业形式联合形成一个复合型的商业空间，使之成为一个新的活力中心。

规划理念
1.以上海原来的城墙理念，将五地块从概念上连接起来（SOHO以及写字楼等）。
　　在A地块拟建一个标志性的塔楼（250 m高），使之成为浦西新地标。
　　SOHO造型创意来自于19世纪海上的交通工具"帆船"，围合为圆弧状（城墙概念），寓意其将成为世界文化的上海之船。
2.充分体现当今世界的节能环保精神。现代化的节能精神与保护自然现状的决心以立体绿带的形式相连，贯穿整个基地。
3.创立一个上海特有的商业中心，A地块因为与上海地铁站的出口相连，将承受巨大的人流量。地铁站出口处建立了一个大空间，使人从意识上尽量忽视地铁的存在。从地铁出来后，首先映入眼帘的将是一个大型的商业空间，从感官上达到忽视地铁的目的。

Nanyou Shopping Park, Nanshan District, Shenzhen City

Planned and Designed by: French AUBE Architecture and Urban Layout Design Company
Project Type: business, office, apartment, hotel and park
Land Area of Project: 126 922㎡
Land Area of Park: 87 615.3㎡
Construction Land Area: 39 306.7㎡
Floor Area: 491 027㎡
Average Floor Area Ratio: 8.48

Base of Nanyou Shopping park project is located in the important urban traffic intersection of Nanshan District, Shenzhen. With the aim of forging a stylish, eco-energy, open and vigorous city parlor, construction will include commercial housing, office, hotels, apartments, public culture areas and urban parks e.t.c.

Concept is firstly coming from the judgment of tension relationship between nature and cities in which there is always a state of wax and wane, mutual erosion. The design adopts an exploding way from center naturally. Ground gushing outside has focused huge energy but been blockaded at the border by the original architectural complex, so it may stretch upwards and finally solidify a risen green landscape, which forms the bowl pattern of lower inside and higher outside. Tails dragging behind the vertical object expand outside maximally, at the same time they absorb the infiltration from urban fabric to land center, which has include the function of business street for daily life.

Here what we do is not to make out the tamed side between human race and nature, but mutual respect between the two.

Planning structure
One horizontal cultural axis: the line to connect the public assembly square, the central eco-park, the schools on west side, the open-air amphitheater and sports parks from the east to west.
Five vertical commercial axes: the north-south commercial pedestrian street located in eight sub-blocks.

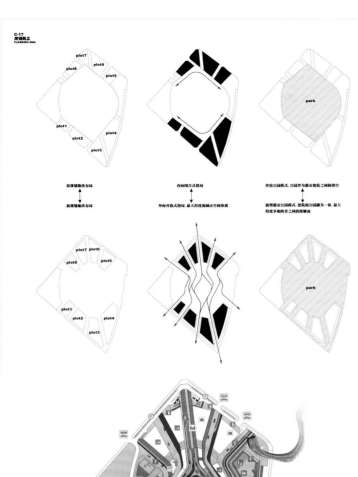

公园景观和建筑融为一体，它们是两个反面全面一理念下的两个反面。景观与建筑之间产生了一种新的地形学的三维模式。它们形成了一个复杂面统一的网络，同时也是一个活跃的系统。它键合了城市的空白区域，也可以对城市的变化做出反应，适应城市发展的需要。

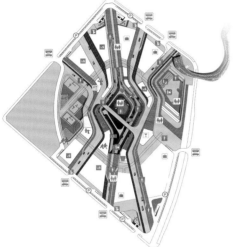
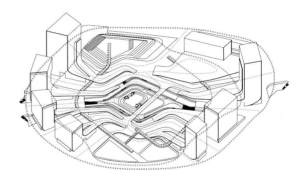

Two commercial concentric rings: commercial pedestrian rings on the ground surface as well as the ones 6 meters above.

One central ecological park: the central water theme ecological park and public underwater art gallery.

There is 1 canter, 2 rings and 6 axes intersecting mutually, stretching to the underground, surface and air, which form an integrated three-dimensional environmental network system.

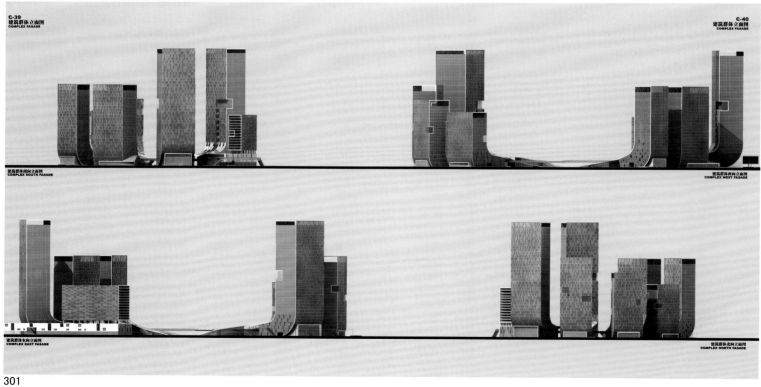

深圳市南山区南油购物公园

规划/建筑设计：法国欧博建筑与城市规划设计公司

项目性质：商业、办公、公寓、酒店、公园。

项目用地面积：126 922㎡

公园用地面积：87 615.3㎡

建设用地面积：39 306.7㎡

总建筑面积：491 027㎡

平均容积率：8.48

南油购物公园项目基地位于深圳市南山区重要的城市交通交叉口，建设内容包括商业、办公、酒店、公寓、公共文化以及城市公园等，目标是打造具有时尚特色、生态节能、开放热烈的都市会客厅。

概念首先出于对自然与城市之间张力关系的判断，二者之间始终存在着此消彼长、相互侵蚀的态势。设计选择了自然由核心"外爆"的方式，向外涌出的地面积聚了巨大的能量，在边界受到既有建筑群的阻滞，进而向上延展，凝固为升起的绿色景观，形成了内低外高的碗状格局。垂

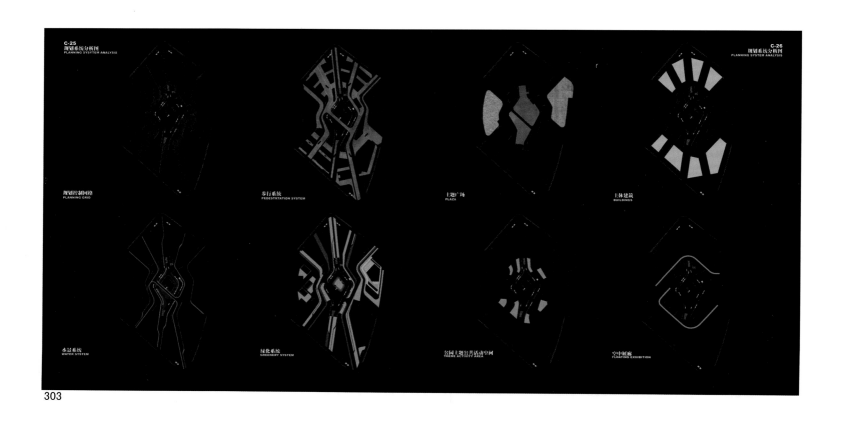

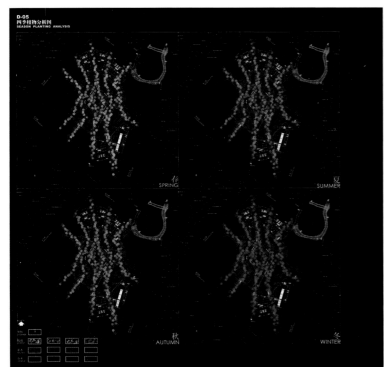

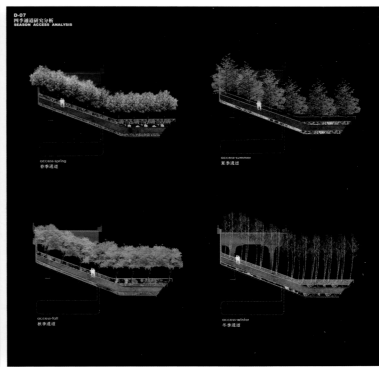

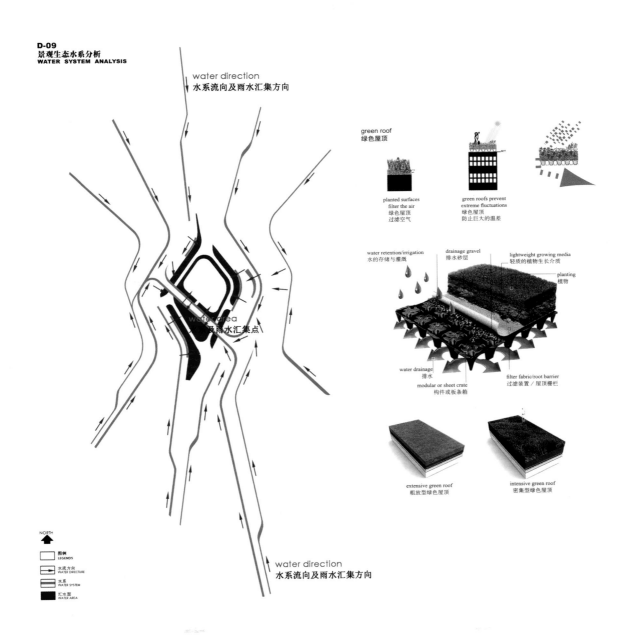

直体后面拖曳的尾翼，在最大限度地向城市外延扩展的同时，吸纳了城市肌理向用地核心的渗透，在日常生活中包容了商业街的功能。

这里要做的不是人与自然之间一方对一方的驯服，而是二者之间的相互尊重。

规划结构

1条横向文化轴：即东西向连接市民集会广场、中央生态公园和西侧学校、露天剧场、运动公园的文化轴线。

5条纵向商业轴：即位于8个分地块之间的南北向商业步行街。

2个同心商业环：即位于地表和地上6米标高的商业步行环线。

1个核心生态园：即中央水主题生态公园和公众水下美术馆。

1心、2环、6轴相互渗透交叉，向地下、地表和空中延绵开去，形成立体化的综合环境网络系统。

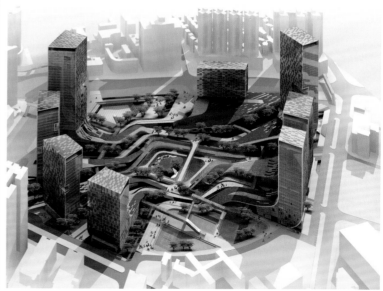